8.50s

GAS LASERS

UNIVERSITY OF CALIFORNIA, BERKELEY
LETTERS AND SCIENCE EXTENSION SERIES

Arnold L. Bloom • Gas Lasers

WILEY SERIES IN PURE AND APPLIED OPTICS

Advisory Editor
STANLEY S. BALLARD University of Florida

Lasers, BELA A. LENGYEL
Ultraviolet Radiation, second edition: LEWIS R. KOLLER
Introduction to Laser Physics, BELA A. LENGYEL
Laser Receivers, MONTE ROSS
The Middle Ultraviolet: its Science and Technology, A.E.S. GREEN, *Editor*
Optical Lasers in Electronics, EARL L. STEELE
Applied Optics, A Guide to Optical System Design/Volume 1, LEO LEVI
Laser Parameter Measurements Handbook, HARRY G. HEARD
Gas Lasers, ARNOLD L. BLOOM

GAS LASERS

ARNOLD L. BLOOM

JOHN WILEY & SONS, INC., NEW YORK · LONDON · SYDNEY

Library of Congress Catalog Card Number: 68-19332
GB 471 08230X
Printed in the United States of America

PREFACE

This book developed from a short course I gave in October 1966 at the University of California Extension in connection with the San Francisco Science Symposium. The course, in turn, was based largely on my review paper, "Gas Lasers," [1] published in *Applied Optics* and *Proceedings of the IEEE*, October 1966. As the material progressed from the review article through the lectures to this book, the emphasis was changed. The review article was intended to present a summary of knowledge and recent results to an audience already thoroughly familiar with the principles of laser operation. The lectures were intended to reach a wider audience, but because of time limitations a basic knowledge of laser physics had to be assumed. Here some material relating to definitions of fundamental quantities and basic laser physics has been added so that the reader will not have to be highly dependent on other texts for background material.

The reader for whom this book is intended, then, is the person who may have access to already-built gas lasers and who wishes to obtain a basic understanding of their operation and use. The book does not contain material on hardware and technology in the sense of giving sufficient detail to enable an inexperienced person to build a gas laser. The reason for taking this approach is that gas lasers with a wide variety of gas species, sizes, and output powers are now available commercially, and a few others can be built quite simply with little or no experience in an average glass-blowing laboratory. Therefore many people who are not interested in building lasers wish to know how to select a laser for their particular needs and what the principal design criteria are that determine limits within which a laser of given specifications can be built.

This book is intended to make it possible for the reader to use the laser intelligently and to provide him with some information on the possible future developments in the field and additional apparatus that can be used with the laser to best advantage.

A problem in the preparation of a book for a highly user-oriented readership is the diversity in educational backgrounds, particularly in mathematics. It is expected that the reader will have an elementary knowledge of atomic physics and of physical optics. Some exposure to laser physics is desirable, and the reader will find in references [2], [3], and [4] in the Bibliography a list of current textbooks that discuss the basic physics of all types of laser in more depth than can be done here. For the mathematics, there are some topics that are essentially mathematical in nature and that have to be treated mathematically, often without adequate physical models. These are attacked with the required mathematical tools as they occur. No great attempt has been made, however, to preserve mathematical rigor and extended derivations have been avoided. Also, the fundamental concepts have been introduced nonmathematically whenever possible. It is hoped, in this way, that the reader who finds some of the mathematics over his head will not find it a hindrance to his understanding of other parts of the book.

For the same reason the use of quantum mechanics to derive basic equations has, in general, been avoided, and dynamical behavior is instead based on an analog of Bloch's phenomenological equations [5] which describe the behavior of classical variables directly. I considered introducing the concept of the density matrix, but I feel that the involved explanation this would require would be distracting and not justified in terms of results. In any event, there is a close relationship between the dynamical variables of Bloch's equations and the elements of the density matrix. For involved theoretical study, particularly of perturbation methods, the density matrix approach is definitely preferred, and the interested reader should consult Bloembergen [6] or Lamb's [7] basic paper.

I have benefited from many discussions with colleagues who are working in the field of gas lasers, but I am particularly indebted to the following persons for contributions that cannot properly be acknowledged by references to published literature: W. E. Bell, J. P. Goldsborough, D. L. Hardwick, D. J. Innes, R. C. Rempel, D. C. Sinclair, and D. L. Wright.

Arnold L. Bloom

Mountain View, California
February 1968

CONTENTS

GAS LASERS

CHAPTER 1

BASIC PRINCIPLES

A. INTRODUCTION

The gas laser is one type of device that is capable of generating coherent electromagnetic radiation at wavelengths shorter than those generally considered to be microwave frequencies. The other devices that have this property employ, as an active medium, crystalline solids, glasses, liquids, or semiconductors. The gas laser, however, has properties that are substantially different from those of the other types of lasers; thus it is desirable to treat them as a separate entity. Throughout this book, "laser" will refer solely to the gas laser.

We shall further restrict our discussion of gas lasers to only those that are excited or "driven" by electric discharge. The reason for this is that discharge-driven lasers are the most common gas lasers, and at present they are the only type available commercially. Experiments have been performed on gas lasers that are driven by optical pumping mechanisms similar to those employed in solid and liquid lasers; however, the experiments have not been pursued and there is little interest at the present time in exploring such lasers further. The reader who is interested in optical pumping of gas lasers can consult texts such as Smith and Sorokin [2] or the first comprehensive review paper of Bennett [8]. Another type of gas laser excitation that is of current research interest is excitation by chemical reaction, either photodissociation or rapid mixing of individual gaseous constituents. This topic, however, is not covered in the present book. In any event, the optical properties of such lasers are essentially identical to those of the discharge lasers, which are discussed in here.

Basically, a gas laser consists of a gas whose atoms or molecules are raised to higher energy states, and an "inversion" exists between a certain pair of energy levels in the gas. To understand the significance of the inversion process, it is desirable to bear in mind the equilibrium conditions that normally exist among the energy levels of the gas that is maintained at a particular absolute temperature T. A gas and temperature equilibrium, whether or not a discharge is maintained through it, has a population distribution among the various energy levels that is described as follows. Let the energy difference between any pair of levels be ΔE; then the ratio of the populations in the two levels is given by

$$\frac{P(E+\Delta E)}{P(E)}=e^{-\Delta E/kT}. \tag{1}$$

In all such cases, for positive values of the temperature T, the upper energy level has less population than the lower energy level. This condition is required by the laws of thermodynamics. If incident radiation having an energy corresponding to the energy difference ΔE between the levels is incident upon the gas, it will be absorbed, the absorption taking place by raising some of the atoms (or molecules)* in the lower state to the upper state. Spontaneous emission processes as well as thermal processes then remove atoms from the upper state back to the lower state and help maintain the temperature equilibrium of the gas.

However, under certain unusual conditions that may exist in an electric discharge, it may be possible to produce a situation in which, between a certain pair of energy levels, the upper level has a greater population than the lower level. This situation is referred to as an "inversion." When this happens, incident radiation, rather than being absorbed by the gas, will stimulate atoms in the upper state to release their energy faster than they would by spontaneous emission and actually contribute more energy to the incident beam than was there originally. This condition is analogous to the gain of an amplifier, and thus the gas discharge medium in a laser can be considered an amplifier of optical radiation. Further details of the physics involved in the laser emission from an inverted population are discussed under the topic of Physics of Gas Laser Operation (pp. 25-29).

As is well known, any oscillator can be turned into an amplifier by providing a suitable type of feedback. In a gas laser the feedback is provided by means of highly reflecting mirrors at either end of an optical path traversing the gas discharge medium. Thus energy that can be employed to start the oscillation begins somewhere in the optical path, traverses the gas, is amplified in the process, and returns eventually to

*The particles in the gas that interact with the laser radiation may be individual atoms, ions, or molecules, depending on the specific laser. In the remainder of the book, we frequently use "atoms" merely as shorthand to refer to whichever type of particle is appropriate.

its original starting point; it is then amplified further, and so on, repeating the process, and this generates the amplification loop that is common to all oscillators. Another feature common to all steady-state oscillators is the fact that some sort of saturation condition must exist in order that the amplified wave does not built up to an infinite value. In a vacuum tube amplifier this saturation condition is provided by the characteristics of the vacuum tube (or other electronic device) that is employed in the circuit. In the gas laser, in fact, in all lasers, the saturation characteristic is provided in the details of the stimulated emission process, which determine in detail how the law of diminishing returns has to apply with respect to injection of additional amounts of energy into the oscillation as the energy content itself grows larger. Eventually, of course, a final limit has to be reached, determined by the rate at which the electric discharge pumps excited atoms into the upper state, which is responsible for the laser inversion itself. A problem in vacuum tube oscillators, which is apparently not present in lasers, may be that of starting the oscillation in the first place. In the laser, the spontaneous emission from the excited state in itself is always sufficient to provide an initial noise signal capable of starting the discharge. As is pointed out in Chapter 3, spontaneous emission noise is always present to some small extent even in a laser operating at high power, and it may be a contributing factor in the noise fluctuations of the laser output.

In summary, then, the gas laser may be thought of as a gaseous medium excited by electric discharge and containing within it a closed optical path in which optical energy can be contained for relatively long periods of time. Because the gain per unit length of most gas lasers is not unduly high, it is desirable in most cases to provide the optical resonator along the path of greatest length through the discharge. Thus a laser in its simplest form will look like Figure 1, which shows the bare essentials that are common to all gas lasers. These are (*a*) a long cylindrical tube containing the gaseous medium, (*b*) a means for exciting the discharge in the medium, and (*c*) a pair of mirrors facing each other,

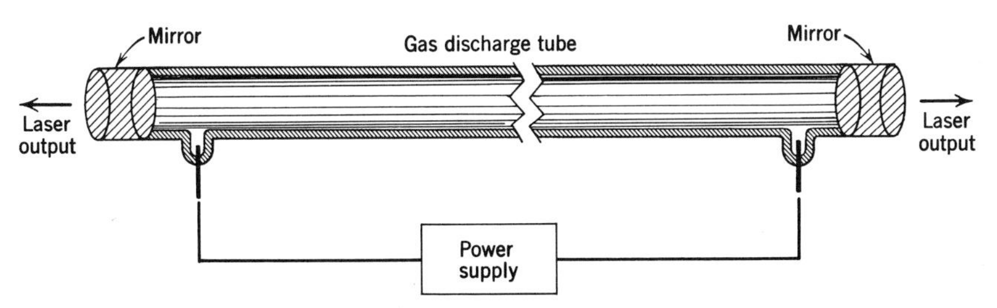

Figure 1. Essentials of a gas laser.

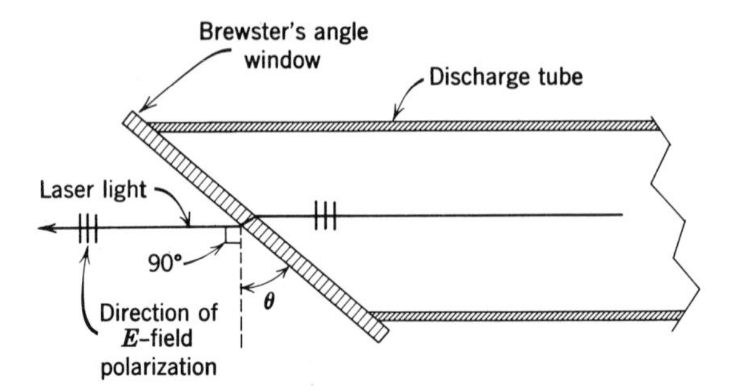

Figure 2. Brewster's angle window.

which constitutes the "resonator" for the laser energy. Often windows of suitable type are inserted between the mirrors and the gaseous medium, though this is not essential to the basic operation of the device. When windows are used, they are generally Brewster angle windows, illustrated in Figure 2. When the angle θ between the perpendicular of the windows and the laser axis is placed so that

$$\tan \theta = n, \tag{2}$$

where n is the index of refraction of the window, no reflection can take place at the surface of the window for the polarization shown in the figure. The other linear polarization suffers a high reflection and therefore gets attenuated too much to provide for laser oscillation, but this does not matter since oscillation in one polarization will effectively remove all of the available energy from the excited atoms provided that coupling exists in the atomic system within the upper and lower state for that particular polarization. In practice this is almost always the case, and the use of Brewster angle windows has the additional advantage that the output is plane-polarized, which is highly useful for many applications. The use of Brewster windows, in general, has been highly successful; Brewster windows when properly manufactured and oriented relative to the laser beam provide a negligible loss of laser energy, generally less than absorption and scattering losses in the mirrors themselves.

Although there is no substitute for actual laboratory handling of a particular type of gas laser in order to understand its properties and characteristics, the following series of photographs may give some indication of the general nature of any given type of gas laser. Figure 3, which was taken with helium-neon lasers, shows the general size and structure of all neutral gas lasers. They are typically of the order of 1 m

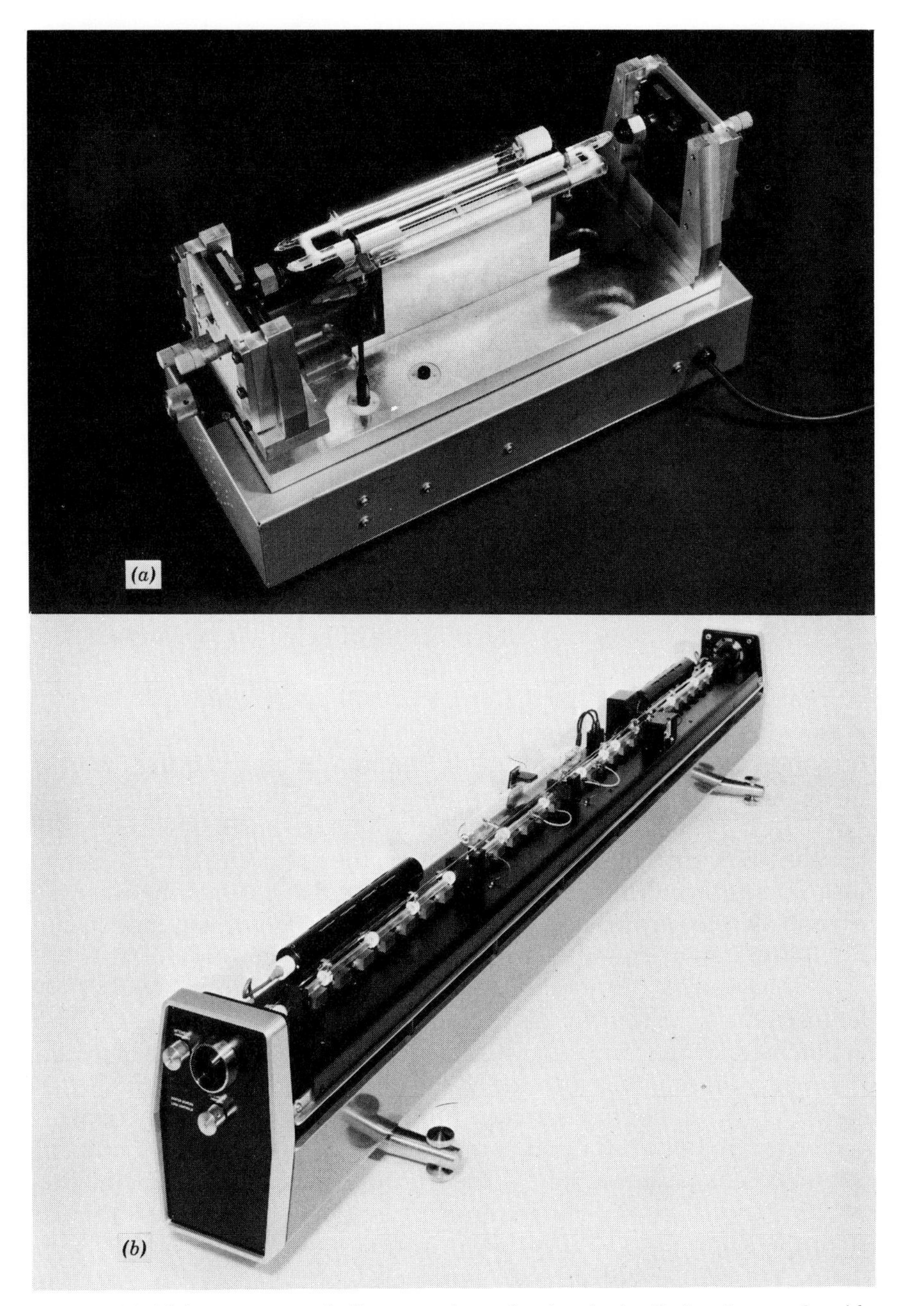

Figure 3. (*a*) A laboratory-setup helium-neon laser showing, in detail, the plasma tube with Brewster angle windows and the resonator mirror mounts. (*b*) A large commercial laser with the cover removed. The rectangular structures under the plasma tube are magnets. (Photographs courtesy of Spectra-Physics, Inc.)

in length but may vary from 15 cm for very short single-wavelength lasers to 2 m or more for high power output. In the helium-neon laser the color of the discharge is somewhat more pale than that of a typical neon sign, owing to the presence of helium, which produces its spectrum lines in approximately the same intensity as those of the neon, which is the lasing material. The pressure at which neutral gas laser discharge tubes operate is somewhat lower than that of rare gas discharge tubes used for other purposes—approximately 1 to 2 torr for the laser versus 10 torr for other types of discharge tubes. The current densities employed are about the same—approximately 100 mA/cm^2 in dc excitation and an equivalent electron density in RF excitation—but all neutral gas lasers are subject to saturation effects if the current density is too high, and in some cases current densities are not as high as could be used merely for the purposes of exciting the emission spectrum. Both RF and dc excitation can be used with this type of laser, dc excitation usually being preferred because of its simplicity and relative lack of electrical interference with nearby laboratory equipment.

Figures 4 and 5 show two different types of noble gas ion lasers. These lasers are characterized by the fact that they must operate at very high current densities, often of the order of 300 to 400 amp/cm^2. For this reason, the plasma itself is at a very high temperature and water cooling or another special cooling means is required to maintain the integrity of the laser structure. The size of the laser structure is typically of the order of 1 m in length, although considerably shorter structures are possible, as shown in Figure 5. However, the requirements for cooling means and the attachments to the laser bore that provide the required current density—either electrodes for dc operation or coupling means for RF operation—also require a considerable breadth in the structure as well as weight. An additional factor contributing to the overall size and weight of the structure is the inclusion of a solenoid for generating an intense magnetic field along the axis of the laser tube, which generally provides a considerable increase in efficiency of the laser operation. Not shown in the photographs are the power supplies that must accompany each laser; since the required power input is often of the order of many kilowatts, this power supply in itself is quite massive and generally requires a special power source (three-phase power, for example) rather than the single-phase 110-V power which is available at most outlets. Thus, although the high-power ion laser does not represent a particularly large installation in terms of length, it does represent a considerable investment in power and in weight.

The method of introducing RF into an RF-driven ion laser is quite different from that of the simple capacitive electrode structures used in

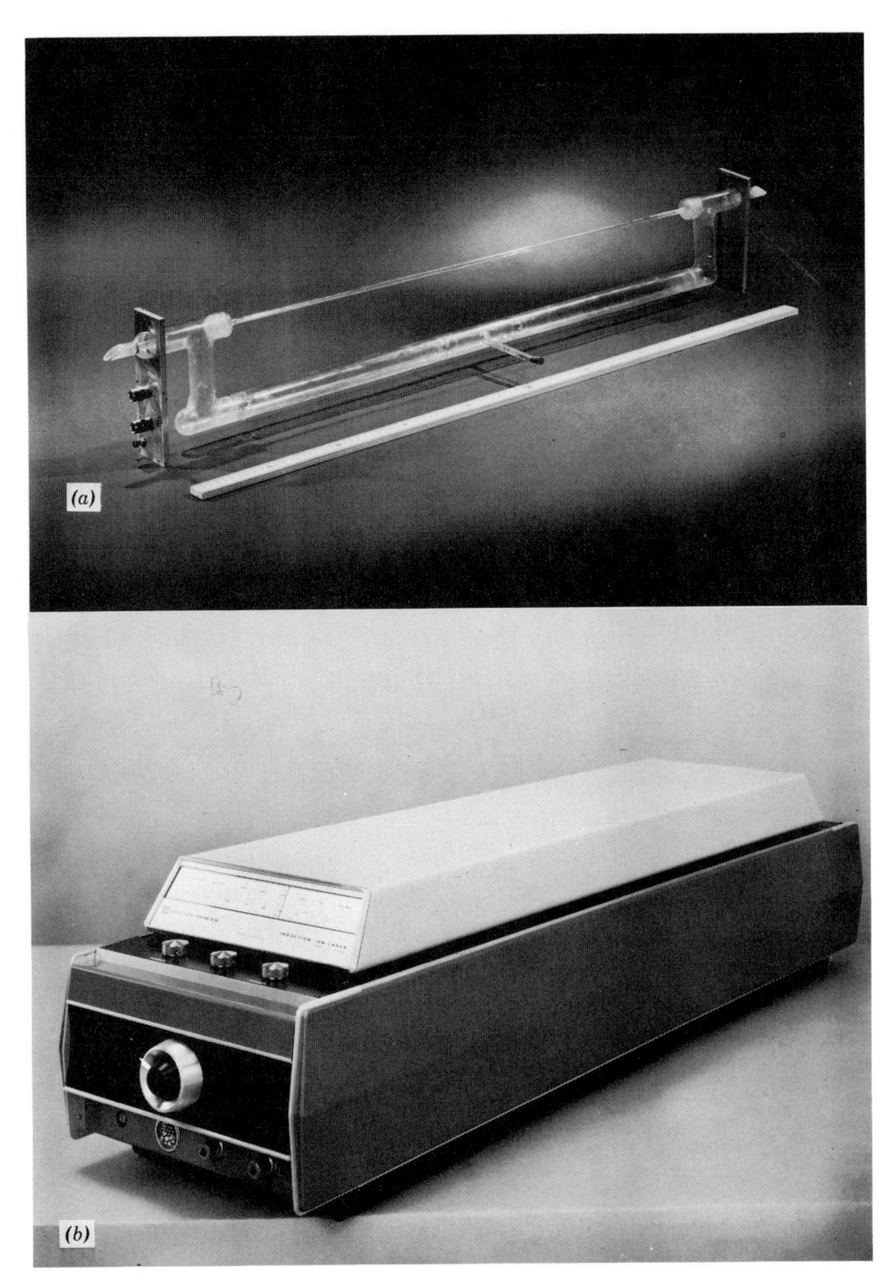

Figure 4. (*a*) Plasma tube of a high-power argon ion laser excited by RF induction. The cooling-water jacket and magnetic field solenoid that generally accompany such a structure are not shown. (*b*) Commercial argon ion laser. The plasma tube in this laser is similar, though not identical, to that of Figure 4*a*. (Photographs courtesy of Spectra-Physics, Inc.)

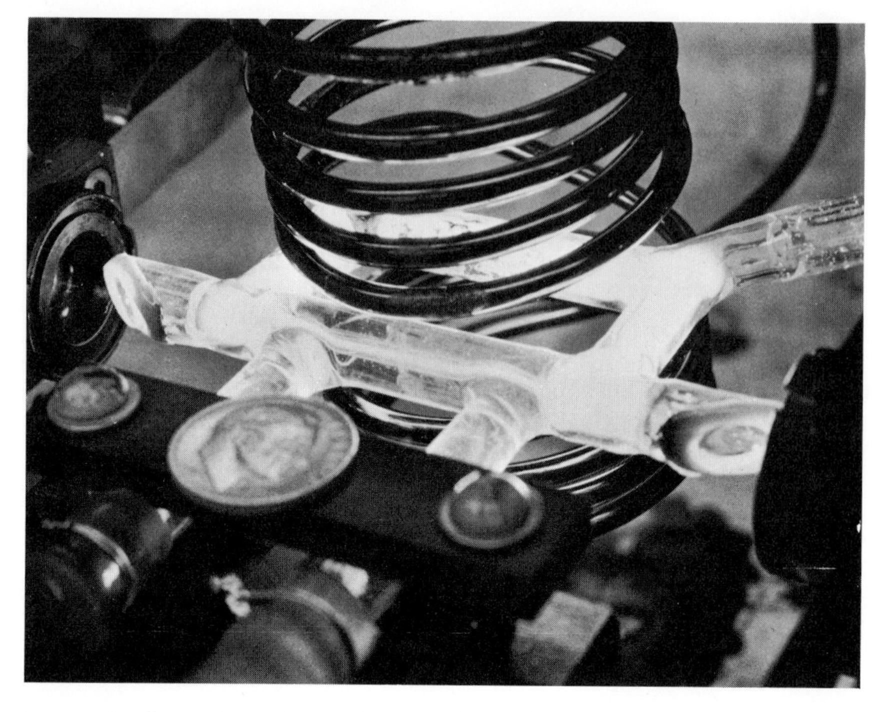

Figure 5. Small experimental argon ion laser, built at Spectra-Physics, Inc., shown with a dime for size reference.

the other types of lasers. Figure 6 shows schematic outlines of both dc- and RF-driven ion lasers, and it will be noted that the RF ion laser has to be coupled magnetically,—as the secondary of a transformer—to the driving RF [9]. This type of coupling is permissible and efficient because the ion laser has such a high current density that the impedance per unit length of the discharge is low and therefore it constitutes a highly conducting current loop. An additional advantage of the RF-excited ion laser, which may possibly be useful in the future, is the fact that absence of any electrodes within the gas permits use of reactive materials as the laser medium. Figure 7 shows those elements that have been observed in CW ion laser action. Of these, chlorine, sulphur, and phosphorus have been known to generate high-power visible output in various parts of the visible spectrum under suitable conditions. These materials cannot possibly be used in a dc-excited ion laser because of their reactive chemical effects on electrodes.

The physical structure of the molecular laser is similar to that of the neutral gas laser, with two exceptions. First, since the wavelengths of

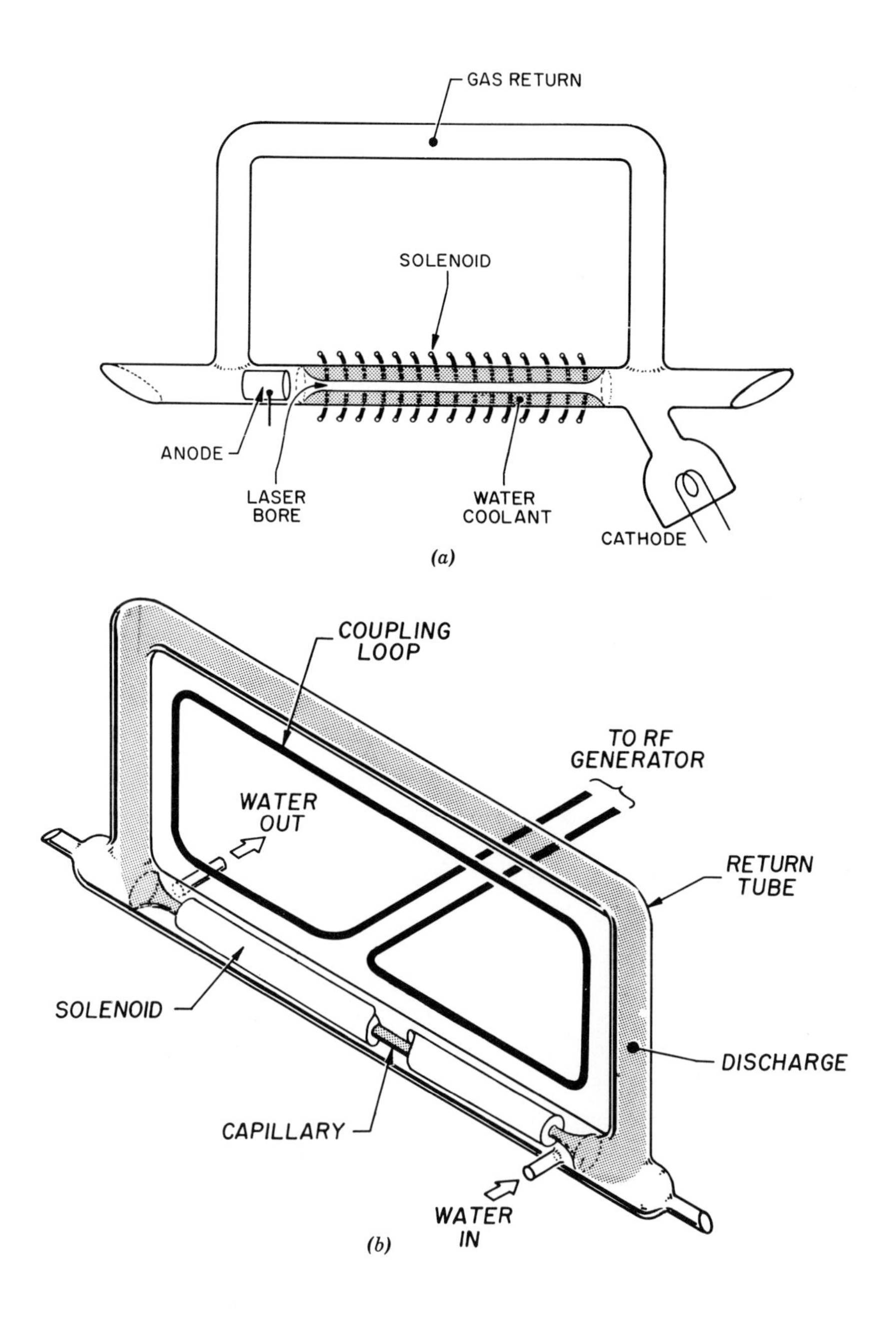

Figure 6. *(a)* Schematic diagram of dc-excited ion laser. *(b)* RF ring excitation of ion laser. (From [1].)

IV	V	VI	VII	VIII	
				He	1
C	N	O	F	**Ne**	2
Si	**P**	**S**	**Cl**	**A**	3
Ge	As	Se	**Br**	**Kr**	4
Sn	Sb	Te	I	**Xe**	5

Figure 7. Elements observed in ion laser action. Helium has not generated ion laser lines, it is included for reference only. Elements in bold type have generated CW laser action. (From [1].)

most molecular lasers are in the far infrared, well out of range of materials that are suitable for Brewster windows, it is almost always necessary to design such lasers with internal mirrors. Second, since the energy separations between transitions in the far infrared are of the same order

Figure 8. High-power CO_2 laser.

of magnitude as thermal energy, it may be necessary to keep the gaseous medium cool in order to prevent upsetting the population inversion from ordinary thermal collisions. This is a particular problem in the CO_2 laser, where it has been shown that the efficiency of operation of the laser depends strongly upon the temperature of the walls of the discharge tube, cold walls being essential for efficient operation. Figure 8 shows a photograph of a high-power CO_2 laser in the laboratory. Note that long length is essential to high power, since it appears that power output is almost directly proportional to length and not to volume. Second, the walls of the discharge are water cooled; cooling with liquid nitrogen or some other refrigerated medium would be even better. The output of this particular laser was of the order of 50 W.

Pulsed lasers—particularly those in which a transient response from the gas is the only type of laser action expected—come in many shapes and sizes but generally have the same appearance as neutral gas lasers. There are two notable exceptions. A hollow cathode type of structure has been used to study ionized mercury, and this appears to have some promise for future applications inasmuch as it has the exceptional ability to operate with a very large laser bore diameter. Figure 9 illustrates this type of laser construction [10]. The other notable structure is one that has been used to obtain a high peak power pulsed output from nitrogen gas [11]. This is shown in Figure 10. It consists of many paral-

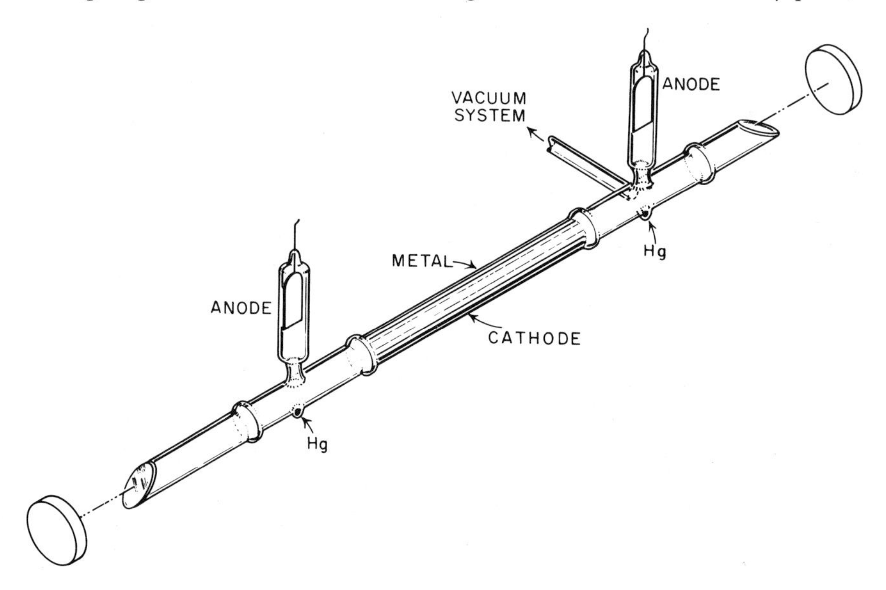

Figure 9. Hollow cathode laser structure for use with mercury vapor. (From [10].)

lel electrodes placed so that the high electric field is across the axis of the tube. By this means, very high fields can be obtained with only "moderate" voltages applied to the electrodes (in at least one experiment, "moderate" was 20 kV). This type of structure probably produces the highest available peak ultraviolet power at the present time.

Other than the particular structures shown in Figures 9 and 10, pulsed lasers usually employ electrodes at either end of the tube similar to that of dc-excited neutral gas or molecular lasers. Excitation is then achieved by discharging a capacitor between the electrodes so that very high peak currents and voltages are available for periods of time of the order of 10 nsec to 1 μsec. RF excitation has been used on occasion, but since a certain number of cycles of the RF are required in order to define the nature of the input voltage, and since the RF is usually of the order of 10 MHz or less, the resultant pulses are relatively long compared to what is usually required for transient laser action, and the output is much more akin to that of a CW laser which is simply pulsed on for a short period of time.

B. DEVELOPMENT OF THE GAS LASER

The first concrete proposal for a gas laser is probably that of the historic paper of Schawlow and Townes [12], published in 1958. This paper, which bases its theory on the then-existing knowledge of stimu-

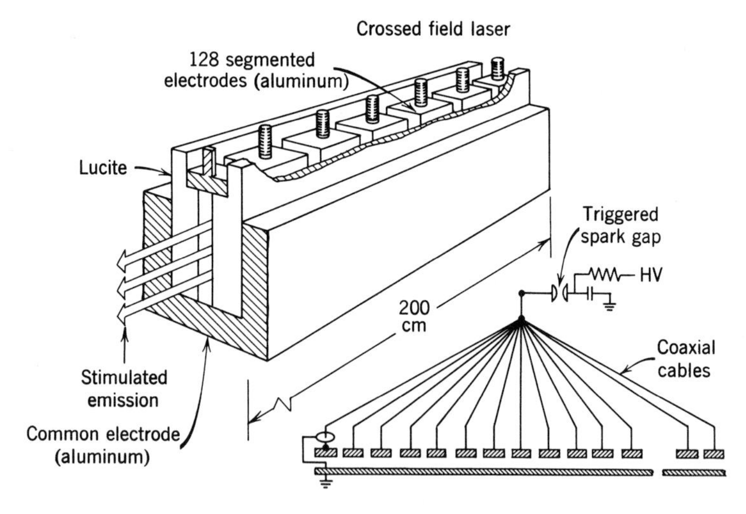

Figure 10. Crossed-field laser structure to obtain very short duration high-voltage pulses. (Reproduced by permission from [11].)

lated emission as related to microwave masers, proposed an optically pumped gas laser as a means of demonstrating the principle of laser action. The paper, considered to be the stimulus for all initial discoveries of lasers of all types, turned out to be surprisingly accurate in predicting the size, wavelength range, and output power for typical gas lasers having moderate input power excitation requirements, though it did not propose the use of a gas discharge for excitation. The proposal for employing a discharge to achieve population inversion and consequent laser action was made by, among others, A. Javan, then of the Bell Telephone Laboratories, about early 1960. Based on a detailed study of excitation processes and lifetimes in excited neon, he predicted possible laser action on the transition $2s_5 \rightarrow 2p_9$ at 1.1177 μ. The first operating gas laser was, in fact, constructed by Dr. Javan and his associates, W. R. Bennett, Jr., and D. Herriott [13]. While this laser did oscillate on the predicted line, 1.1177 μ, it was found that much stronger oscillation was obtained on adjacent lines in the vicinity of 1.15 to 1.2 μ. In fact, the 1.1177 μ line has very rarely been seen from a helium-neon laser other than the first one; the author has observed it in some of his experiments, and undoubtedly it has been observed by others as well, but it is known to be a very difficult line to excite.

This point is made in regard to the relation of theories of population inversion to the actually observed results. The history of the discovery of gas laser transitions has been one of extremely poor agreement between theories and subsequent experimental discoveries. The aforementioned example is one of the few in which a theory was actually able to predict the existence of a subsequently observed laser transition; another particularly successful one was the prediction by C.K.N. Patel [14] of the enhancement of CO_2 laser action at 10.6 μ through interaction with nitrogen. For the most part, however, the theoretical approach has not been the one that has resulted in the discovery of the many existing laser transitions. Most of these transitions were discovered by the simple expedient of preparing a laser structure according to the scheme of Figure 1, filling it with some gas at pressures of the order of 1 torr, employing mirrors that were highly reflective within some wavelength range, and "looking" to see if any laser light came out of the structure.

On the other hand, there is one area of laser research in which theory has been particularly helpful—in fact, essential—to the further development of the technique. This is the theory of optical resonator structure and modes. The original paper of Schawlow and Townes pointed out that the concept of modes as applied to a microwave cavity would not be appropriate in the optical region, and it suggested that the resonator should perhaps consist of a Fabry-Perot type of structure, that is, a pair of flat mirrors enclosing the laser medium and directed toward each other. The exact nature of the modes of this structure was left open to question, and at least some investigators in the field felt that a "mode" of such a resonator might correspond to the "rings" or fringes that are observed when one looks at an incoherent light source through a Fabry-Perot interferometer. This partially misconceived notion persisted throughout the early work until the actual discovery of the gas laser, and the author can vouch for the fact that this misinterpretation of the true nature of an optical resonator mode provoked some lively discussions at scientific meetings around 1960.

It was not until after the discovery of the gas laser that the correct theory of the optical resonator began to emerge. Fox and Li [15] employed numerical integration on a computer to study energy storage in a plane-parallel resonator, and subsequently Boyd and Gordon [16] and Boyd and Kogelnik [17] did corresponding calculations for resonators employing curved mirrors. The results of this work form the basis for a good part of Chapter 3 of this book. Its importance in promoting the development of gas lasers after the initial discovery cannot be underestimated. Until this theoretical work was done, the gas laser was at best a marginally operating device whose oscillation depended upon almost impossibly precise tolerances in the end mirror adjustments. The theoretical studies on curved mirror resonators showed that resonators could be devised that were relatively insensitive to mirror adjustment and whose intrinsic losses could be much lower than those of a plane-parallel resonator, allowing observation of laser action in media with much lower gain than was heretofore thought possible. The plane-parallel resonator has since almost dropped out of existence for practical laser work, and all discoveries of new gas laser transitions have been performed with curved mirror resonators.

The next major experimental step was the discovery of visible laser output at 6328 Å by White and Rigden [18] in the spring of 1962. Besides providing a great stimulus for applications, the visible laser made possible a much greater understanding of the problems involved in optimizing laser structures and in demonstrating precisely what the difference was between a coherent and an incoherent beam of light. The work also stimulated additional research on other possible laser transitions and many more were found in neutral atom lasers whose structures were akin to that of the helium-neon laser, but all of these transitions were in the infrared. Such experiments continued for a period of about two years, during which considerable engineering development was done on the helium-neon lasers.

In 1963 the significance of employing pulsed operation to obtain laser transitions that could not be observed continuously was demonstrated by several investigators. Mathias and his co-workers experimented with molecular nitrogen and carbon monoxide, obtaining visible transitions from carbon monoxide [19]. However, the most important of these experiments was probably the discovery by Bell in the fall of 1963 of pulsed ion laser action in mercury vapor [20]. The discovery that laser action could be obtained from the ionic spectra of atoms suggested a whole new range of possibilities for laser operation in various materials, and a large number of investigators immediately shifted their efforts to the study of the ion lasers. An indication of the extent to which this occurred is the fact that the discovery of the argon ion laser, the most important type of ion laser, was made independently and almost simultaneously in at least four laboratories around the world [21]. Observation that argon and other noble gases could operate as ion lasers continuously as well as pulsed was made soon thereafter [22].

By using a combination of the successful techniques—weak glow discharges such as the helium-neon laser, pulsed discharges, and high current discharges such as are used in the ion lasers—investigators have come up with a total of approximately 1000 laser transitions in various gases. Although the rate at which new lines are being discovered has slowed down considerably in the last year or two, additional lines are still reported from time to time in the scientific journals.

The most recent important development in the laser field is probably the discovery that the CO_2 laser at 10.6 μ is capable of operation at undreamed-of power and efficiency. The present field of laser research appears to be aimed in two general directions: (a) efforts to raise the power and efficiency of existing types of laser, and (b) attempts to obtain a deeper knowledge of the basic physics of such lasers as the helium-neon lasers. Particularly to be noted in this connection is work on subtleties such as details of the line shape in lasers and the effects of weak magnetic fields on their operation. On the other hand, there is no reason not to expect additional important discoveries comparable to that of the CO_2 laser to occur in the near future.

The development of the gas laser up to the present time has suggested a broad classification of all lasers into four general types. This classification, which will be used freely in the text, and the general characteristics of lasers within each category, are as follows:

1. *Neutral atom lasers.* Lasers employing relatively weak discharges and having moderate power output and gain, the laser transition spectra being that of neutral atoms in the discharge.
2. *Ion lasers.* Lasers operating on transitions in ionized atoms in the

discharge. These lasers generally operate at high current densities, high power inputs, and occasionally with relatively high power output.

3. *Molecular lasers.* Lasers somewhat similar to the neutral atom lasers but employing molecular spectra. They employ moderate current densities in the discharge but can occasionally be operated with large volumes and hence high input power. Many but not all lasers of this class emit in the far infrared.

4. *Pulsed lasers.* A heterogeneous group that overlaps into the other three classes but which includes many lasers that cannot be made to operate continuously for a number of reasons. Many lasers of this class have relatively high peak power output and gain, but because of the required pulsing conditions their average power is not particularly high.

The detailed characteristics of specific lasers in the first three classes are discussed in Chapter 2.

C. FUNDAMENTAL PROPERTIES AND DEFINITIONS

We define here terms that are used throughout the book to describe properties that are fundamental to the operation of any gas laser.

1. Output Wavelength

All gas lasers have the property, not shared to any great extent with the other types of laser, that the wavelengths of laser emission are precisely defined and occur only at certain very discrete parts of the spectrum; they are not "tunable" to any extent. Appendix A lists the wavelengths, in microns, of a number of representative gas laser transitions. All output wavelengths of all gas lasers occur at positions represented by spectrum lines in the emission spectra of the particular gas used in the laser. However, experience has shown that the spectrum lines that are observed in laser emission are very rarely the strongest lines seen in spontaneous emission; in fact, a number of laser lines represented the first experimental observation of the spectrum line involved. The reason for this lack of coincidence between laser transitions and strong spectrum lines is inherent in the physics of laser operation; laser operation demands the generation of a population inversion or "piling up" of an atomic population in the upper of the two energy levels that are involved in the laser transition. On the other hand, a strong spectrum line is strong precisely because there is a high probability that an atom that is deposited in the upper energy level will decay spontaneously at a rapid rate to the lower energy level.

A word should be said here regarding the notations used for the energy levels that give rise to laser transitions. A complete list of such levels for atoms and ions is given in the work of C. E. Moore-Sitterly [23], now regarded by most investigators in the field as standard. We shall not have much occasion to use energy level notation in this book, but when it is used it will generally follow the notation of Moore. The helium-neon laser has, by convention, come to be described in a somewhat empirical notation known as the "Paschen" notation. This notation was developed before there was a complete understanding of the atomic structure of the neon atom, and it is used because it is simple to write and in any event possesses a 1 to 1 correspondence with the actual energy levels of the neon atom. For present purposes, it should be considered merely a code for designating the neon energy levels.

2. Modes—Spatial and Temporal

The end mirrors of a laser define a resonator, and as in all resonators there exist resonances or normal modes, field configurations that represent energy storage within the resonator and are self-sustaining within the resonator for relatively long periods of time. The modes of an optical resonator are similar to those of a microwave resonator in that they represent configurations of the electromagnetic field that are determined by boundary conditions. Here, however, much of the similarity ends. A microwave resonator typically is a box whose dimensions are defined by the coordinates x, y, and z, and a mode may be defined by the symbol TEM_{mnq} where the numbers m, n, and q define the number of nodes in the x, y, z directions, respectively, of the standing wave within the resonator. A similar notation can be adopted for the optical resonator of a laser, with the z direction defined as the axis of the laser, that is, the line joining the two end mirrors. However, in a laser resonator mode TEM_{mnq} the number q defines the number of half-wavelengths of light between the two end mirrors and typically is a large number, of the order of 10^6. The numbers m and n, on the other hand, are typically very small, on the order of 0, 1, or 2. The situation is illustrated graphically in Figure 11. Because of the essential asymmetry between the axes xy and the axis z, and the large disparity in mode numbers, the properties of a mode configuration in the x, y directions are largely independent

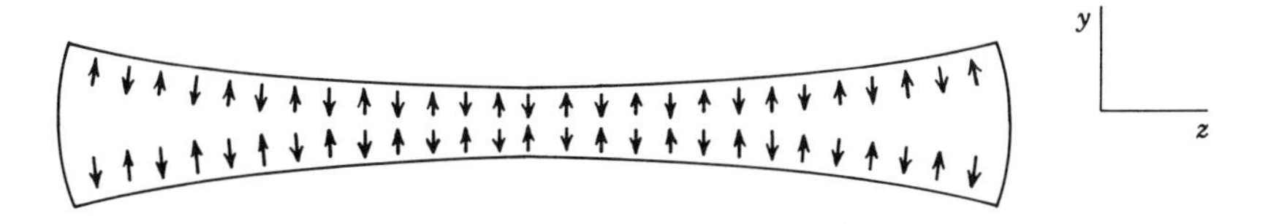

Figure 11. A schematic "snapshot" of the E-vector in a laser resonator mode at a particular instant of time. The example is shown for mode number $n = 1$. The mode number q is, in practice, much larger than is indicated here.

of the configuration in the z direction, and vice versa. Thus it is common to speak of the configuration in the x, y directions as the spatial or transverse mode configuration, and, quite independently, to speak of the z direction as the longitudinal or temporal mode configuration. The reason for the word "temporal" will become clear shortly. For simplicity, we shall immediately drop the use of the word "configuration" when speaking of the resonator modes.

3. Doppler Width

A property fundamental to all spectra emitted by gas discharges is the shifting in frequency or wavelength caused by the motions of the emitting atoms. Although thermal motion takes place in three dimensions, only motion parallel to the line of sight is important in causing Doppler shifts, transverse motion causing only an exceedingly small second-order effect. If the motion of the atoms in the discharge is completely Maxwellian—completely thermal in nature—then the intensity distribution caused by the Doppler effect is given by

$$I(\nu+\Delta\nu)=I_0 \exp\left(\frac{-Mc^2\,\Delta\nu^2}{2\nu^2kT}\right) = I_0 \exp\left(\frac{-\Delta\nu^2}{\delta^2}\right) \quad (3)$$

where

$$\delta=\frac{2kT\nu}{Mc^2}. \quad (4)$$

Here ν is the unshifted emission frequency, $\Delta\nu$ the amount of shift, I_0 the central intensity of the Doppler-broadened line, M the atomic or molecular mass, c the speed of light, and kT the thermal energy at the temperature T. We shall define δ to be the Doppler width of the line, that is, the frequency spread in which the line intensity varies from peak to $1/e$ of its peak intensity. It is customary in most spectroscopic literature to define the Doppler width as the spread between the two half power points of the distribution. This quantity is $1.6651 \times \delta$, as defined above.

The Doppler width of the emission line that gives rise to laser action is of fundamental importance in determining the operating characteristics of the laser. First, the gain of a laser is determined by, among other factors, the number of atoms in the laser that are capable of interacting with radiation having a specified frequency. If the total number of excited atoms in the laser have their interaction frequencies spread out over a range, then clearly there are fewer atoms available to interact at any given frequency. Second, the Doppler width determines a range of frequencies over which laser operation can actually occur, within any given laser transition. As an approximate rule, the intensity of laser action at some frequency ν is proportional to that given by (3). The primary factors that enter into this distribution and the width δ are the mass M and the temperature T. This assumes, of course, that the distribution is in fact Maxwellian with a well-defined temperature T. One would expect in some lasers, particularly the ion lasers, that other effects such as acceleration by fields would modify the distribution to something other than Maxwellian, but in the cases that have been investigated it appears that this is not so [24]; rather, a well-defined temperature T can be defined in all cases.

4. Frequency Separation between Modes

If a laser is operating in one spatial mode but in more than one temporal mode, then the different temporal modes must be at different frequencies. This is the reason for use of the word "temporal" in this connection. The frequencies at which temporal modes can occur are set by the boundary conditions at the end mirrors, which determine that laser oscillation can occur only when there are an integral number of half wavelengths between the mirrors. In a mode denoted by the notation TEM_{mnq}, the number q denotes the nodes in the electric field between the end mirrors. The number of half wavelengths in the resonator is then either q or $q + 1$, depending in detail on the spatial mode involved (a question that is immaterial at this point). If q' is an integer that denotes either q or $q + 1$, as appropriate, then the wavelength of the laser oscillation is given by

$$\lambda = \frac{2L}{q'} \tag{5}$$

and the frequency is therefore given by

$$\nu = \frac{cq'}{2L}. \tag{6}$$

The frequency separation between modes corresponding to q' and $q' \pm 1$ is thus given simply by

$$\Delta\nu = \frac{c}{2L}, \tag{7}$$

where c is the mean speed of light in the laser medium and L the length of the resonator (in fact L is an "effective" length taking into account possible wavelength changes because of refractive index effects). The quantity $c/2L$ is thus fundamental to many details of laser operation. It is also, incidentally, the inverse of the time required for a light signal to travel a complete round trip through the laser starting at any given point and bouncing between the two mirrors.

Although the quantity $c/2L$ can be calculated easily enough, it is important to have a "feel" for the magnitude of the $c/2L$ frequency spacing in the following discussions. Table 1 gives the $c/2L$ frequency spacing for typical resonator lengths used in gas lasers.

When the laser is operating in several spatial modes as well as in several temporal modes, then the frequency allocations of the various modes become a much more complicated matter. This is discussed in Chapter 3.

5. Gain, Loss, and Useful Output

The laser, like any classical resonant oscillator, can have its operation characterized in terms of the gain of the medium, the internal losses, and the useful output.

The gain of the medium can best be considered by treating it as an untuned amplifier. Such an amplifier would consist of the laser medium alone in the discharge tube, without the end mirrors; thus one could insert an optical beam or "signal" at one end and receive an amplified signal from the other end. Consider a weak signal inserted into a laser amplifier at some frequency ν in the vicinity of a laser transition. If the gain of the laser is moderate, the increase in signal intensity as a function of ν, as ν is varied, will simply follow the Doppler curve given by (3). [Strictly speaking, the differential gain per unit length obeys (3); the total gain throughout the length of the laser tube constitutes the integral of the gain over the length, which does not differ significantly from (3) unless the gain is very high, in which case it has a sharper central peak than (3) would indicate.] In any event, the maximum gain for a weak signal, inserted into a laser amplifier, occurs at the center of the Doppler curve. This amount of gain is generally known as the *peak unsaturated gain,* and it is this quantity we refer to whenever we speak of "gain," unless otherwise specified. If a laser is in oscillating condition, and one attempts to amplify an external signal through the laser medium simultaneously, then the gain for this signal is defined as the *saturated gain,* and this is generally a considerably smaller number than the unsaturated gain. The difference between unsaturated and saturated gain is important primarily in discussions regarding competition effects between modes in lasers that are operating simultaneously in several modes.

The losses in a laser are any media that remove energy from the laser resonator. Typical sources of loss are: transmission of energy through the end mirrors, scattering by imperfections in the mirror surface, scattering and reflections from imperfect Brewster windows, and diffraction losses within the resonator. The useful output of the laser is usually taken by transmission through one of the end mirrors. If both end mirrors are transmitting to some extent, then both transmissions may be considered useful output, or one may be considered useful output and the other a source of internal loss, depending on the application of the laser.

TABLE 1

Typical $c/2L$ Frequency Spacings for Lasers of Moderate Length

Length L (cm)	$c/2L$ (MHz)
10	1500
30	500
60	250
100	150
180	83

The general rule for stable operation of all lasers is this: The peak unsaturated gain of the laser medium must be greater than the sum of the losses and the useful output. This is elaborated upon in the section on Steady State Equations of Laser Operation.

6. "Hole Burning"

This term, which will be used frequently in the following discussions, is intended to describe graphically the process whereby energy for the laser operation is extracted from only certain groups of atoms occupying various positions under the Doppler profile. When a laser is oscillating, it is drawing energy out of those atoms that are at resonance with whatever modes are being sustained within the laser, and in the process of doing so the amount of population inversion which gives rise to the laser energy is reduced: the population is "drained" of its energy. Another way of stating this is to say that the laser mode reduces the saturated gain in the vicinity of its oscillation frequency. Consider, for simplicity, a laser operating in only one temporal mode. It will interact with the atoms whose positions under the Doppler profile lie within a certain range of the oscillation frequency. (How wide this range is depends in detail upon the atomic interactions and is discussed later.) If this range is appreciably narrower than the Doppler width itself, then the saturated gain is depressed in the region immediately around the oscillation frequency but is unperturbed elsewhere in the Doppler profile. We have thus "burned a hole" in the Doppler profile. If several temporal modes are operating, then, of course, holes are burned in the vicinity of each oscillation frequency.

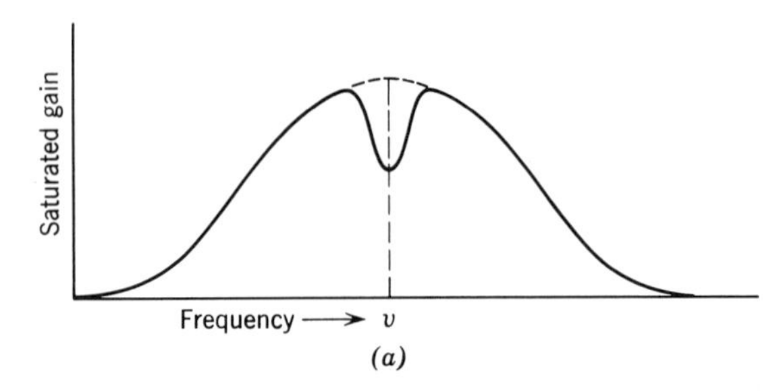

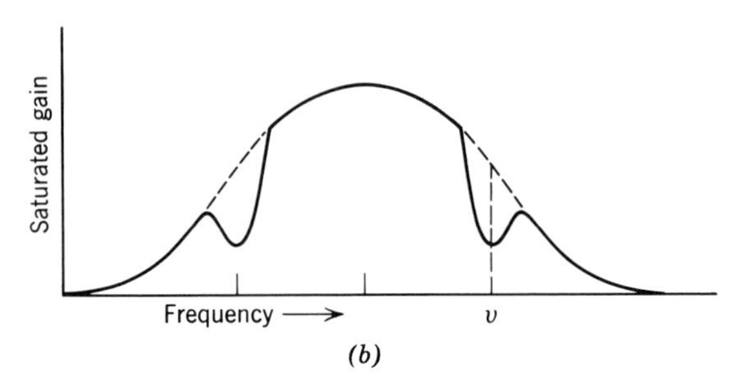

Figure 12. Hole burning by a single temporal mode. (*a*) Mode frequency ν centered within the Doppler profile. (*b*) Mode frequency offset from Doppler center.

Figure 12 shows the process of hole burning for a single temporal mode, for the case where the mode frequency is at the center of the Doppler profile and where it is offset to one side. When it is offset, two holes are burned, one at the oscillation frequency and one at a position symmetrically located with respect to the center of the Doppler curve. The existence of the two symmetrically located holes has to do with the existence of a standing wave within the laser resonator. This standing wave can be considered to be made up of two traveling waves moving in opposite directions, thus there is symmetry within the laser with regard to motion in opposite directions along the axis. If there is a hole burned among atoms traveling at a velocity v in one direction along the laser axis, it is because those atoms are at resonance, owing to a Doppler shift, with one of the traveling waves of the laser mode. However, because of the symmetry, this also means that atoms traveling at a velocity $-v$ will be at resonance with the other traveling wave. Hence there exist two symmetrically located holes. If the oscillating laser mode is at the center of the Doppler curve, then clearly only one hole can be burned instead of two, and this might be expected to cause a difference in output power under some conditions. The phenomenon does in fact exist and is discussed in Chapter 3 under the topic of the power dip. Also, if a resonator is devised to use only traveling waves instead of standing waves, then clearly the symmetry with respect to Doppler line center no longer exists and only one hole is burned for any given oscillating mode. A short discussion of traveling wave resonators is given at the end of Chapter 3.

7. Homogeneous and Inhomogeneous Broadening

These terms relate to details of laser operation as discussed in the next section. However, since they are closely related to the question of hole burning, it is convenient to define the terms here. If a particular laser operates in such a way that the Doppler gain curve is uniformly "burned", that is, uniformly and thoroughly saturated, then it is said to be homogeneously broadened. If, on the other hand, it operates so that one or more discrete holes are burned leaving sections of unsaturated gain, then it is said to be inhomogeneously broadened. Homogeneous broadening can occur for one of two possible reasons: (a) the Doppler width is narrower than the theoretical width of the hole that can be burned (this occurs primarily in far infrared lasers), or (b) the laser is operating in many temporal modes spaced closely enough together so that no region of the gain curve is left unaffected by a temporal mode. Whether a laser is homogeneously or inhomogeneously broadened is therefore a question of design in some cases. Since inhomogeneous broadening represents the existence of untapped sources of laser power, as indicated by unsaturated portions of the gain curve, it represents less efficient operation of the laser than is the case with homogeneous broadening.

8. Near-Field and Far-Field Optical Patterns

The terms "near field" and "far field" occur frequently in the literature on lasers. We shall have occasion to use them in connection with discussions on laser spatial modes and on the propagation of the laser beam after it leaves the source. For the benefit of readers who have not previously encountered these terms, we give short definitions of them here.

When a beam of light is output from a laser or any other optical device, it has some sort of well-defined energy distribution as a function of position across the beam cross section. This is the near-field intensity distribution across the beam. The beam is propagated into space initially according to the laws of geometrical optics, but after a while diffraction effects tend to modify the intensity distribution in the beam. Eventually, a point may be reached where the beam has spread out to such a degree that it covers a considerably larger area than would have been predicted by geometrical optics, and the intensity distribution in the beam is determined entirely by diffraction. This is the far-field distribution. More precise definitions in terms of transform properties are given in textbooks on physical optics; however, the preceding definitions will be adequate for our purposes.*

Our definitions are intended, in a sense, to apply to plane or approximately plane wave propagation, although they also apply to propagation in the form of spherical wave fronts. In particular, consider the near-field distribution to be in a converging spherical wave. According to geometrical optics, such a wave front should converge to an infinitely small point at the focus; however, we know that diffraction effects will always prevent the distribution at the focus from being infinitely small. It follows therefore that the distribution at the focal point of a converging wave is always the far field distribution of the initial converging wave. A much more rigorous proof of this statement is given by Born and Wolf [25]. This fact is important in connection with the theory of laser resonators, as well as the treatment of propagation of laser beams given in Chapter 4 of this book.

There are two near-field intensity distributions whose far-field distributions will be of particular interest to us. One is the intensity distribution of a uniformly illuminated wave passed through a circular aperture of radius a. The far-field pattern of this distribution is given by

$$I(v)=I_0\left(\frac{2J_1(v)}{v}\right)^2, \tag{8}$$

where $J_1(v)$ is the Bessel function of order one and v is a length coordinate whose actual measurement depends on the wavelength of the light and the details of the optical system involved (it is defined more exactly in Chapter 4). The distribution of (8) is known as the Airy distribution and, when observed on a screen, is often known as the Airy disk. (The energy distribution in the Airy disk is shown in Figure 43.) Another near field distribution that will be of interest to us is the Gaussian distribution given by

$$I(r)=\frac{I_0e^{-r^2/\alpha^2}}{\pi\alpha^2}. \tag{9}$$

The far-field distribution corresponding to (9) is given by

$$I(v)=\frac{I_0}{\pi}e^{-v^2}. \tag{10}$$

*In textbooks on physical optics, the diffraction pattern that gives rise to the far-field intensity distribution is usually defined as the Fraunhofer diffraction pattern.

In other words, for the Gaussian distribution the far-field distribution has the same form as the near-field distribution. This fact makes the Gaussian distribution particularly important in discussions on the generation and propagation of laser modes.

D. PHYSICS OF GAS LASER OPERATION

The following discussion briefly outlines the topic, emphasizing specific points that are essential for the understanding of all gas lasers. For much more detailed discussions, the reader is referred to other texts that discuss laser physics in general [2-4] and to the earlier review articles on gas lasers such as those of Bennett [8,26], which discuss excitation and inversion mechanisms of gas lasers in much more detail that is possible here.

1. Stimulated Emission versus Spontaneous Emission

The basis for the physics of all maser and laser operation is the existence of stimulated emission from an upper energy level of a quantum mechanical system. Consider an atom having two energy levels separated by energy E, where $E = h\nu$. The atom is surrounded by an electromagnetic field which can be considered to be made up of waves traveling in various directions, and each of these waves in turn can be considered to be made up of a number, n_ν, of quanta or photons, each of energy $h\nu$. If the atom is in the lower state, then the probability that the atom will absorb one of these quanta and arrive in the upper state is

$$p\,d\Omega = \frac{8\pi^3 e^2 \nu^3 d\Omega}{hc^3} \,|\,\mu\,|^2\, n_\nu \cos^2\theta. \tag{11}$$

The probability that an atom in the upper state will emit a photon and drop to the lower state is

$$p\,d\Omega = \frac{8\pi^3 e^2 \nu^3 d\Omega}{hc^3} \,|\,\mu\,|^2\, (n_\nu + 1) \cos^2\theta. \tag{12}$$

In these formulas p is the transition probability per unit time, $d\Omega$ is an element of solid angle, ν is the optical frequency, μ is the matrix element for the transition, θ is an angle relating the polarization of the field to a "preferred" polarization of the atom, and e, h, and c have their usual meanings as fundamental constants. These formulas are derived in most textbooks on quantum mechanics; the form used here was taken from Heitler [27]. For our present purposes, the term n_ν can be considered approximately equal to the number of photons existing in each mode of the laser resonator, if the active medium referred to is in the resonator. The terms involving n_ν in (11) and (12) thus represent the absorption and stimulated emission probabilities, respectively, whereas the extra term—"1"—in (12) represents the spontaneous emission.

Because it does not depend on preexisting conditions in the electromagnetic field, in the spontaneous emission the phase of the output is completely unpredictable, only the *rate* at which the spontaneous emission occurs is predictable. Thus consider an ensemble of atoms capable of radiating spontaneously, such as those in the upper laser level of a gas laser. Although many atoms are emitting spontaneously at the same time, the fact that the phases are random means that mean-square averaging can be applied to the output of the ensemble taken as a whole; all phase relationships and interferences between the radiation of different atoms average out and the spontaneous radiation can be treated statistically merely as the sum of the radiation of the individual atoms. As a result it is isotropic in space and has the characteristics of "white noise" within the Doppler bandwidth of the emission. Spontaneous emission is always present in a laser therefore and represents a source of noise as well as a source of signal useful for starting the laser oscillation. However, only that part of the spontaneous emission which is within the solid angle subtended by the laser mode actually becomes part of the stored energy in the laser mode.

The stimulated absorption and emission, on the other hand, have the important property that they are coherent in phase with the electromagnetic field variations which produce them. The reason that this is so of course is buried in the fundamentals of quantum electrodynamics, but some insight can be obtained by going back to the original treatment of thermodynamic equilibrium in the radiation field by Einstein [28]. This theory, which preceded quantum mechanics by quite a few years, showed that the coefficients of stimulated absorption and emission have to be equal and opposite in nature, so that radiating matter can achieve thermodynamic equilibrium with the radiation field. Now, consider the stimulated absorption from the standpoint of a dipole or other multipole interacting with the radiation field. In order that energy be absorbed it is essential that there be a consistent phase relationship between the phase of the electromagnetic field and the phase of the oscillating dipole that characterizes the absorbing particle. If the absorbing particle is considered as an oscillating dipole, it must also produce its own radiation field, and the phase of the radiation field must be such as to subtract, by destructive interference, from the intensity of the incident field at large distances from the particle. If a consistent phase relationship did not exist, energy would not, on the average, be absorbed from the radiation field. The difference between absorption and stimulated emission, according to theories of radiation equilibrium, is only that the phase relationship is reversed, so that in stimulated emission the particle now does work on the radiation field and adds constructively to the field by interference at long distances from the radiation source.

The coherent nature of stimulated emission has an important corollary: it determines that *spatially* the stimulated emission must be identical to the incident radiation. This is a consequence of the spatial distribution of an ensemble of emitting particles in the incident wave, not necessarily a property intrinsic to the stimulated emission itself. A single particle, considered analogous to a radiating electromagnetic dipole, has a high probability of radiating energy in many directions. However, when an ensemble of particles is spatially distributed through a propagating wave, then the phase relationships must be considered. With regard to propagation of a free wave, Huygens' principle states that propagation of a traveling wave takes place as if each point of the wavefront consisted of sources that are generating wavelets in phase with the incoming wave. If we now take an ensemble of particles emitting in phase, and assume that they are dense enough, then Huygens' principle must also apply to the stimulated emission of these particles, so that their emission actually becomes part of the incoming wave and amplifies it without changing it spatially. Thus an active medium spatially distributed in a laser resonator with well-defined oscillating modes can be expected to radiate its stimulated emission only into those modes. The spontaneous emission of these particles, being bound by no phase relationships whatever, is free to be emitted in all directions.

2. Gain per Unit Length

It is often useful to be able to estimate the gain of the laser medium in terms of the atomic parameters. The original paper of Schawlow and Townes [12] calculates the conditions required for stable laser operation in a derivation that requires a knowledge of the number of electromagnetic modes in a "box." Following is a somewhat simpler derivation based upon a treatment of resonant absorption given by Mitchell and Zemansky [29]. Such a treatment is valid because, as we have seen, there is a close relationship between stimulated emission and resonant absorption.

The simplest case of resonant absorption is that in which a beam of light at frequency ν is propagated through a resonant medium with a unique absorption constant. As the light propagates through the medium, its intensity varies with distance x, according to

$$I = I_0 e^{\alpha x} \tag{13}$$

where α, if negative, is the absorption constant. If α is positive in (13), then clearly we have gain and the quantity α is the gain per unit length in the medium.

The calculation can be performed in terms of the Einstein A and B coefficients for thermodynamic equilbrium, as defined by Mitchell and Zemansky. Here A is the probability per unit time that an atom in the upper state will spontaneously emit a photon and drop to the lower state, and Bn_ν is the probability that an atom is stimulated to absorb or emit a photon, as the case may be, into a mode of the electromagnetic field containing n_ν photons. [There is thus a close relationship between the A and B coefficients and the quantities defined by (11) and (12). For simplicity, we assume that the statistical weight factors of the upper and lower states are equal. If they are not, a minor change in the formulation is required, as given by Mitchell and Zemansky.]

The relationship between the A and B coefficients is

$$\frac{A}{B}=\frac{2h\nu^3}{c^2}. \tag{14}$$

In order that the concept of a transition probability be valid, it is necessary to assume that the radiation field and absorption resonance are not infinitely sharp but cover a small frequency range $\Delta\nu$. We suppose that there are ΔN_ν atoms per cubic centimeter in the lower state and $\Delta N'_\nu$ in the upper state resonant over the frequency range $\Delta\nu$. Mitchell and Zemansky then show that

$$-(\alpha\Delta\nu)=\frac{h\nu}{4\pi}B\,(\Delta N_\nu-\Delta N'_\nu). \tag{15}$$

This could also be written as

$$\alpha=\frac{\dfrac{h\nu}{4\pi}B\,(N'-N)}{\Delta\nu}, \tag{16}$$

expressing the fact that, for given *total* populations N' and N in the upper and lower states, α is inversely proportional to the frequency range, a fact which is frequently noted elsewhere in this book. Note also that α is positive and there is net gain instead of absorption whenever N' is greater than N, that is, whenever there is a population "inversion."

Returning now to (14), we note that the A coefficient is the inverse of the lifetime of the upper state against spontaneous emission to this particular lower state. If we call this lifetime τ and substitute in (14) and (16), we arrive at the following equation for the gain coefficient as a function of the lifetime τ:

$$\alpha=\frac{\left(\dfrac{c^2}{8\pi\nu^2}\right)(N'-N)}{\Delta\nu}. \tag{17}$$

In this formula N and N' must have the units of total population *per unit length* in order that α have the correct units of inverse length. This formula is fundamental and applies to all lasers provided the appropriate quantity τ can be measured. It is sometimes convenient to express α in terms of the matrix element of the transition. If the matrix element is μ, then the well known relationship between μ and τ is

$$\tau = \frac{3hc^3}{64\pi^4\nu^3\mu^2}. \tag{18}$$

We have then

$$\alpha = \frac{\frac{8}{3}\left(\frac{\pi^2\nu\mu^2}{hc}\right)(N' - N)}{\Delta\nu}. \tag{19}$$

In many lasers the total gain αx is small enough that the approximation

$$e^{\alpha x} \approx 1 + \alpha x \tag{20}$$

can be used. In these cases it has become convenient to speak of the gain in terms of a fractional increase in signal; thus one speaks of the gain as "percentage gain" where the percentage is $\alpha x \times 100$. Otherwise, a logarithmic unit such as decibel is more useful and is applicable to all situations. Moreover, if the gain is expressed in logarithmic units, then, if taken as a function of frequency under the Doppler profile, it is always proportional to the ordinate of the Doppler profile. For small gains, the quantity $(e^{\alpha x} - 1)$ is also approximately proportional to the ordinate of the profile, and this is the situation usually assumed in considering output power of a laser as a function of frequency deviation from the center of the Doppler curve. On the other hand, if the gain is high and one considers power output, significant deviations from the actual Doppler shape can be expected. Mitchell and Zemansky discuss the flattening effect of a large absorption coefficient upon the observed profile. Where there is gain instead of absorption, the effect is to narrow the profile. If the natural linewidth* is very narrow, then the half power point on the Doppler Gaussian becomes, in fact, the square-root point for the total gain, so that, if the gain is very high, the observed profile may be very narrow indeed. A limit is eventually set by the fact that the observed profile cannot be narrower than the natural linewidth. Use of this phenomenon has been made to measure natural linewidths of laser transitions [30].

*The natural linewidth is the range of frequencies over which atoms having identical velocities will be in resonance. To a good approximation, it is the sum of the A coefficients of the upper and lower states.

3. The Bloch Equations

A description of the dynamical behavior of the atoms interacting with the laser radiation fields is perhaps obtained most easily by use of equations derived by Bloch [5] for use in nuclear magnetic resonance. It may seem surprising that a theory intended for use with nuclear magnetic resonance should have application in laser theory, but this is so due to a general unity in form imposed by quantum mechanics. It may be desirable to go briefly into the reasons for this. The original analysis of nuclear magnetic resonance concerned a system involving only two energy states. In a laser, the atoms involved in laser action have more than two energy levels, but as a rule only two of these levels—the upper and lower laser states—are actually involved in the dynamical behavior of the system to first approximation. Thus in both cases we have two-level systems subject to perturbation by resonant applied fields and to restoration of equilibrium conditions by relaxation processes. Now, a fundamental axiom of quantum mechanics is that dynamical variables of a system are associated with operators on the wave functions of this system. For a system consisting of only two energy levels, such a wave function might take the form

$$\psi = \begin{pmatrix} a_1 \\ a_2 \end{pmatrix} \tag{21}$$

where the wave function is a two-element vector and a_1 and a_2 are the probability amplitudes for finding a particle in states 1 and 2, respectively. The operators on a two-element vector are four-element matrices, and it can readily be shown that there can be only four linearly independent 2×2 matrices, including the identity matrix. (The Pauli spin operators are an example of a set of such matrices.) Thus there can be only four independent dynamical variables of a two-level quantum mechanical system, and if the variables are specified for one such system, the variables in any other system will be isomorphic to them in some sense. Since four such dynamical variables are defined by the Bloch equations, they can be taken over directly for use in a laser system. The remaining question is the extent to which the relaxation processes described by Bloch can be employed in describing lasers. These relaxation processes, describing simple exponential approaches to thermal equilibrium, are probably somewhat simpler than those that actually exist in lasers, but the approximation is good enough to make the equations useful for most purposes. For examples of the fruitfulness of this approach the reader is referred to Bloembergen [6] and to work on the transient response of laser systems to very sharp pulses, a few examples of which are given in the bibliography [31-33]. Many other examples also exist in the literature.

In the Bloch formulation, the identity matrix corresponds to the total available population; for present purposes it can be assumed to be constant and ignored. The remaining three dynamical variables are denoted by u, v, and M_z, whose corresponding operators are as follows:

$$u \approx \begin{pmatrix} 0 & 1 \\ 1 & 0 \end{pmatrix} \cos \omega t + \begin{pmatrix} 0 & -i \\ i & 0 \end{pmatrix} \sin \omega t, \tag{22}$$

$$v \approx -\begin{pmatrix} 0 & 1 \\ 1 & 0 \end{pmatrix} \sin \omega t + \begin{pmatrix} 0 & -i \\ i & 0 \end{pmatrix} \cos \omega t, \tag{23}$$

$$Mz \approx \begin{pmatrix} 1 & 0 \\ 0 & -1 \end{pmatrix}. \tag{24}$$

Here, $\omega = 2\pi\nu$ and ν is a frequency which is generally chosen to be equal to the frequency of the "driving" field that generates the resonant behavior. Note that this is not necessarily the resonant frequency of the atoms themselves; like any other resonant system, an atomic resonance can be driven slightly off-frequency, though with reduced response.

The physical significance of the dynamical variables is as follows: M_z corresponds to the *saturated* population difference between the upper and lower laser levels; v corresponds to an oscillating dipole moment which is either plus or minus 90° out of phase with the driving electromagnetic field. It can be shown easily that this dynamical variable interacts directly with the energy content of the driving field, either doing work upon the field, as in a laser, or absorbing energy from the field in an absorption line. Whether it absorbs or emits energy depends, of course, upon the sign of v. The remaining variable u is the in-phase component of oscillating dipole moment. This component, because of its phase relationship, cannot change the energy of the driving field but instead changes the phase velocity for propagating waves. It is responsible for anomalous dispersion in absorption lines and for similar effects in lasers. However, our primary concerns here are the variables v and M_z.

It is also necessary to define the relaxation terms to be used with these variables. Rather than use the original Bloch notation of relaxation times T_1 and T_2, we shall employ a notation more in accord with spectroscopic practice and define relaxation *rates* γ_{ab} and γ'_{ab}. γ_{ab} is the *spontaneous emission* or *hard collision* relaxation rate for the laser transition. It represents the average rate at which atoms interacting with the applied field are removed from the interaction, either by spontaneous emission processes or by hard collisions that remove the atoms to entirely different states. The quantity γ'_{ab} includes not only the relaxation rate γ_{ab} but also additional terms caused by soft collisions, which may cause the phase of the wave function of an interacting atom to wander but not completely destroy its coherence. More detailed discussions of the meanings of these terms have been given by other authors [34], and it is also not obvious where one draws the line between "hard" and "soft" collisions, but we are concerned here only with the basic essentials of the theory.

Within the framework of these definitions, then, the differential equations governing the dynamical variables u, v, and M_z are

$$\frac{du}{dt} + \gamma'_{ab}\, u + \Delta\omega\, v = 0 \tag{25}$$

$$\frac{dv}{dt} + \gamma'_{ab} v - \Delta\omega\, u = 0 \tag{26}$$

$$\frac{dM_z}{dt} + \gamma_{ab} M_z - \omega_1 v = (N' - N)\,\gamma_{ab}. \tag{27}$$

Here, the following additional definitions are needed: $\Delta\omega = 2\pi(\nu_0 - \nu)$, where ν_0 is the natural resonant frequency and ν is the driving frequency. ω_1 is defined by

$$\omega_1 = \frac{|\mu| E}{\hbar} \tag{28}$$

where μ is the matrix element and E the mean-square value of the driving field; this is the strength of the atomic interaction. N' - N is, of course, the *unsaturated* or initial value of population difference as used in (16) to (19).

There are numerous transient (time-dependent) solutions of these equations that have interesting properties; these have been discussed by Bloch [5], Torrey [35], Hahn [36], and many others. We shall be more concerned with the steady state solutions. These, obtained by solving (25) to (27) with the differentials set equal to zero, are

$$u = \frac{\omega_1\, \Delta\omega\, (N' - N)\, L(\Delta\omega)}{(\gamma'_{ab})^2} \tag{29}$$

$$v = \frac{-\omega_1 (N' - N)\, L(\Delta\omega)}{\gamma'_{ab}}, \tag{30}$$

$$M_z = \left[1 + \left(\frac{\Delta\omega}{\gamma'_{ab}}\right)^2\right] (N' - N)\, L(\Delta\omega), \tag{31}$$

where $L(\Delta\omega)$ is a general resonance expression common to all three of these formulas and is given by

$$L(\Delta\omega) = \frac{1}{1 + \left(\dfrac{\Delta\omega}{\gamma'_{ab}}\right)^2 + \dfrac{\omega_1^2}{\gamma_{ab}\gamma'_{ab}}}. \tag{32}$$

The quantity M_z, as we have pointed out, is directly proportional to the saturated population difference for a homogeneous population whose resonant frequency is ν_0. The energy actually injected into the laser resonance mode by the laser medium is proportional to $\nu \mid E \times v \mid$; since we are generally concerned with extremely small variations in ν we shall consider ν to be constant. For further development of these equations, we wish to express energy injection in terms of the energy already present in the laser, which we define as W. W is proportional to E^2. By substituting W into (28), (30), and (32) and collecting all constants, we find this simple expression for the rate of energy injection:

$$\frac{dW}{dt} \propto \nu \mid Ev \mid \propto \frac{GW}{1+\left(\frac{\Delta\omega}{\gamma'_{ab}}\right)^2+SW}, \tag{33}$$

where G is the unsaturated gain and S will be referred to as the saturation constant.

4. Steady State Equations of Laser Operation

We are now in a position to calculate the power output of a gas laser as a function of its gain, losses, and mirror transmission. However, it becomes necessary at this point to differentiate between homogeneously and inhomogeneously broadened lasers. The Bloch equations [(29) to (32)] can be used directly for the homogeneously broadened case, because they apply to ensembles of atoms whose resonances are all essentially at the same frequency. We shall derive the steady state conditions for homogeneous broadening first, going on to the inhomogeneous case later. It should be borne in mind that "steady state" applies to the output of almost all gas lasers including pulsed lasers, because the relaxation times involved in reaching dynamic equilibrium are generally shorter than the pulse length, even when submicrosecond length pulses are used.

The increase in energy in a laser resonator due to the active medium is proportional to the v term, or to (33). With the driving field at frequency ν_0, and reduced to simplest terms, the energy increase can be written as follows:

$$\frac{dW}{dt}=\frac{GW}{1+SW}. \tag{34}$$

The removal of energy from the resonator is due to the output mirror transmittance T and undesirable losses A. Thus for the losses we have

$$\frac{dW}{dt}=-(A+T)\,W. \tag{35}$$

At steady state, the energy increase equals the energy removal so that we have

$$(A+T)\,W=\frac{GW}{1+SW}. \tag{36}$$

If the driving field is not at frequency ν_0, then, referring to (33), it will be seen that the effect on (34) is merely that of changing the proportionality constants G and S, but the form of the equation itself is unchanged.

The output power P of the laser is then:

$$P=TW=\frac{T}{S}\left(\frac{G}{A+T}-1\right). \tag{37}$$

Negative values of P clearly have no meaning and thus the equation states the obvious—that steady oscillation is obtained only when the gain is greater than the loss. Note that the saturation constant S appears only as a proportionality term. From (32) it will be seen that S is determined by relaxation rates and by the matrix element of the transition; thus in a given laser its value is out of the control of the experimenter. Its numerical value can be determined by experiment if this seems desirable.

Equation 37 forms the basis for calculating the mirror transmittance that gives optimum output power for fixed gain and loss. This optimum transmittance is

$$T_{opt}=\sqrt{GA}-A. \tag{38}$$

This relationship is similar to but more complicated than the well known relationship in electric circuits that states that optimum efficiency is obtained when the load equals the internal resistance. It applies, strictly speaking, to the case of homogeneous broadening with moderate gain—of the order of not more than 40 or 50% per pass. For higher gain one must take into account the exponential growth of the oscillation signal, whereas (38) merely assumes a lumped active element emitting power into the resonator mode. Rigrod [37] has given expressions for the case of arbitrary gain, and has plotted the values of T_{opt} for various gains and losses. Even these formulas, however, do not take into account the effect of spontaneous emission. All of the aforementioned formulas—either (37) or Rigrod's formulas—predict zero power output if the transmittance of one mirror is equal to unity (perfect transparency) because then, in principle, there is no longer any laser resonator. In fact, under conditions of sufficiently high gain it is possible for the spontaneous emission from one end of the laser to be amplified sufficiently to saturate

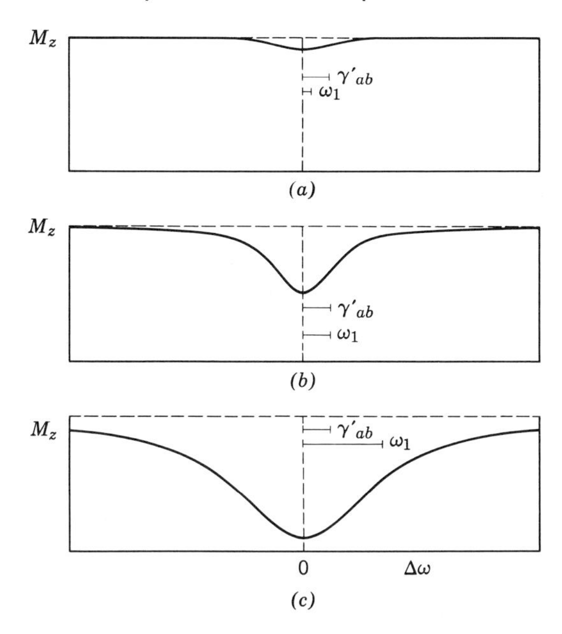

Figure 13. Plot of M_z as a function of $\Delta\omega$ and ω_1. (*a*) ω_1 small, the hole in M_z has a depth proportional to ω_1 and a halfwidth approximately γ'_{ab}. (*c*) ω_1 large, the hole has a constant depth and a halfwidth of approximately ω_1. (*b*) An intermediate case.

the active medium at the other end of the laser (or at the same end if there is a mirror at the opposite end). This phenomenon has been given the name of *superradiance**. It has been observed numerous times in lasers of high gain, particularly pulsed lasers where very high gains per pass can be achieved for short periods of time. Because of the absence of a structure that defines modes, however, a superradiant laser does not generate light having a very high degree of coherence.

The calculation of steady state conditions for an inhomogeneously broadened line differs from that for homogeneous broadening in that one must consider the existence of a driving field at fixed frequency ν and an ensemble of particles covering a range of values of $\Delta\omega$. At each of these values of $\Delta\omega$, there is a value of M_z given by (31) that is determined by substituting for $\Delta\omega$ in that equation. Figure 13 shows M_z as a function of $\Delta\omega$ for several intensities of the driving field. This is quite

*The term *superradiant state* was originally coined by R. H. Dicke [38] to describe certain coherent radiating states in the active medium. Its use for the purpose described here is thus a misnomer; however, it has become so common in the literature that we shall use it to mean amplified spontaneous emission.

obviously the hole-burning phenomenon described earlier. Note that for very strong driving fields the hole becomes quite wide, enabling the field to draw energy from atoms that are quite far removed from the resonant frequency.

For inhomogeneous broadening, the quantity v must be replaced by an integral of the contributions to v from atoms covering the entire range of resonant frequencies. Such a computation, applied to steady state laser conditions, was first derived by J. P. Gordon at the Bell Telephone Laboratories. Because the computation is in the form of unpublished notes not generally available, however, we shall derive it independently here. Clearly, the integral must take the form

$$v_{inh} = \int_{-\infty}^{\infty} v(\nu_0 - \nu)\, F(\nu_0)\, d\nu_0, \tag{39}$$

where $(\nu_0 - \nu) \equiv \Delta\omega/2\pi$ as before. The integral is thus of the general form

$$v_{inh} = \int_{-\infty}^{\infty} \frac{v_0\, F(\nu_0)\, d\nu_0}{(1+SW) + [(\nu_0 - \nu)/\gamma_{ab}']^2}, \tag{40}$$

where $v_0 \propto v(\Delta\omega = 0)$. The function F represents the distribution of atoms over the frequency range, for example, the Gaussian curve of the Doppler profile. If F is taken to be the Gaussian, the integral can be solved but not in terms of elementary functions [39]. A simpler distribution, which is nevertheless physically meaningful, is a Lorentzian distribution centered on the driving frequency ν, given by

$$F(\nu_0) = \frac{1}{1 + [(\nu_0 - \nu)/\delta']^2}, \tag{41}$$

where δ' is the half-width at half-maximum of the distribution. Substitution of (41) into (40) gives an integral that can be solved analytically. The result is

$$v_{inh} = \frac{\gamma_{ab}'\, v_0\, \pi}{1 + \dfrac{1+SW}{\delta^2}} \left(\frac{1}{\sqrt{1+SW}} - \frac{1}{\delta} \right), \tag{42}$$

where $\delta = \delta'/\gamma_{ab}'$. There are two limiting cases. For δ very small, (42) reduces to the same form as the result for homogeneous broadening. This corresponds to a hole width larger than the distribution width. In the other limit—δ very large—the result is

$$v_{inh} = \frac{\gamma'_{ab} v_0 \pi}{\sqrt{1 + SW}}. \tag{43}$$

This is the result obtained by Gordon. We can now write it in the form

$$\frac{dW}{dt} = \frac{GW}{\sqrt{1+SW}}, \tag{44}$$

and, combining this with (35), we obtain the following for the power output of an inhomogeneously broadened laser:

$$P = \frac{T}{S}\left[\left(\frac{G}{A+T}\right)^2 - 1\right]. \tag{45}$$

The important fact about (45) is that under some conditions the power varies with the square of the gain rather than with of the first power, as in (37). This can be seen qualitatively in terms of the hole burning concept, since higher field intensities within the laser not only result in higher output power but also in a wider hole being burned so that more atoms are able to emit their energy into the laser mode. This relationship is very important in certain types of lasers that exhibit inhomogeneous broadening characteristics, particularly the CW ion lasers.

There is no simple expression for the value of T_{opt} in the inhomogeneously broadened case. Rigrod [37] has calculated T_{opt} for homogeneous and inhomogeneous broadening and for various gains and losses. The general form of these results is shown in Figure 14. Figure 14*a* shows the general nature of the power output curve as the transmittance T is varied from 0 to 100%. Figure 14*b* shows the power output at T_{opt} as a function of the losses A in the resonator for both the homogeneously and inhomogeneously broadened cases. The following important difference should be noted. Most lasers operate with relatively small losses, corresponding to operation close to the Y axis of Figure 14*b*. Under such conditions, if the losses are reduced still further, for example, by a factor of 10, the power output that can be obtained at the new value of optimum transmittance is changed hardly at all for homogeneous broadening. However, in the case of the inhomogeneously broadened line, the new value of T_{opt} is smaller, thus permitting less leakage of energy out of the resonator and building up a greater intensity in the standing wave. This in turn increases the interaction of the resonator mode with the atoms in the inhomogeneously broadened line and results in a considerable increase in the available power. If (24) is taken literally, this curve increases without limit as it approaches the

Y axis. This, of course, is not the case in practice, since a leveling off will occur at the point at which the "hole" becomes equal to the width of the Doppler curve. However, the dramatic increase in available output power from an inhomogeneously broadened laser as the losses are reduced has been observed experimentally and is of particular importance in small, single frequency lasers where the ratio of gain to fixed losses is not overly large.

5. Zeeman Effect

The Zeeman effect is a well known spectroscopic effect involving splitting of atomic spectrum lines when a magnetic field is applied. It enters into the present discussion only indirectly, because magnetic fields are sometimes applied to lasers for various reasons, with consequent effects on the gain and output linewidth. The following is intended as the briefest possible introduction; for more detail the reader may consult any textbook on atomic physics or spectroscopy.

The Zeeman effect in its simplest form is the splitting of a spectrum line into several components with well-defined polarizations, the splitting being generally proportional to the magnetic field. Figure 15 shows

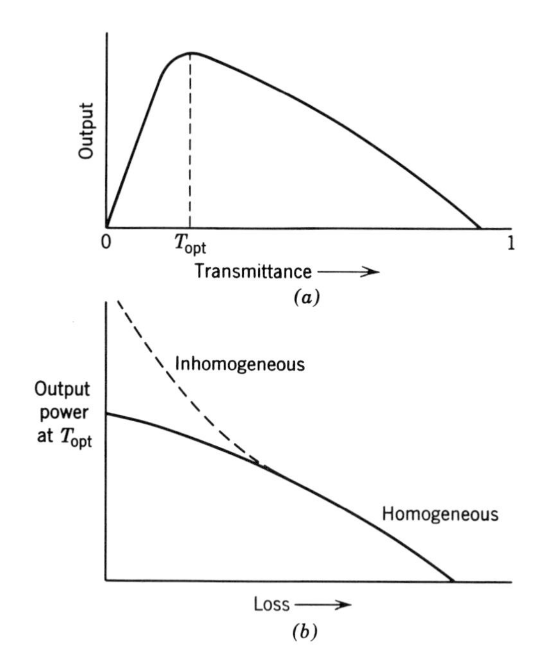

Figure 14. Typical curves of output power as a function of mirror transmittance and loss.

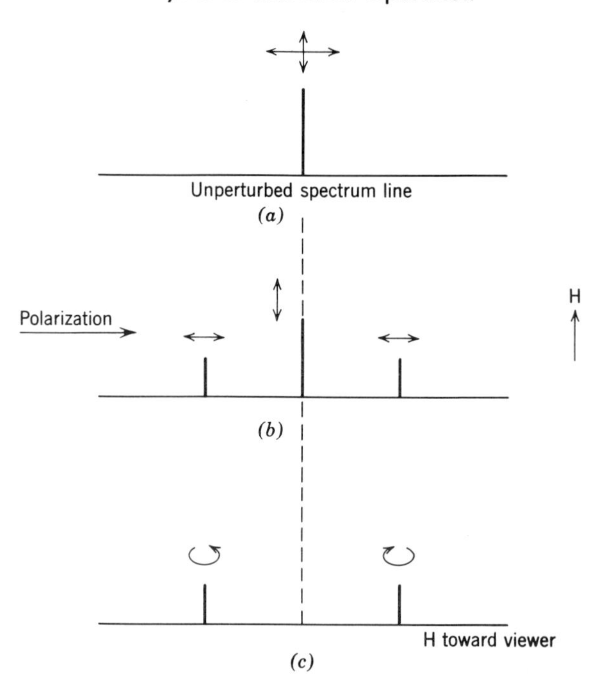

Figure 15. Example of Zeeman effect. (*a*) Unperturbed spectrum line. (*b*) Field perpendicular to line of sight. (*c*) Field parallel to line of sight.

the components and polarizations observed when the magnetic field is parallel and perpendicular, respectively, to the line of sight. It should be emphasized that Figure 15 shows only one simple case of Zeeman effect; in many specific examples each of the polarization components shown is further split into other components. The amount that the outer components are displaced from original line center can vary widely for different spectrum lines but is generally in the vicinity of 1.4 MHz/G of applied field. In the case of a field perpendicular to the line of sight the spectrum line is split into three linearly polarized components; with a longitudinal field there are two circularly polarized components whose senses of polarization are opposite to each other.

In a laser, application of a strong magnetic field will not, in general, split the line into distinct spectral components unless the splitting is appreciably greater than the Doppler width. If the splitting is only comparable to the Doppler width, an effective broadening of the line occurs, with consequent reduction of gain in any given frequency range and with the gain somewhat stronger in preferred polarizations than in others. Strong longitudinal magnetic fields are particularly important for efficient operation in the CW ion lasers, and they can be expected to provide appreciable Zeeman splitting there. A signal of arbitrary polarization injected into one end of an ion laser can be considered decomposed into two circularly polarized components, one of which will be amplified preferentially over the other by the laser medium because it may fall within one or the other of the Zeeman-split components. If Brewster angle windows are present, however, they treat the amplified laser radiation as made up of two linearly polarized components, one of which is partially reflected out of the cavity. Thus, in establishing static equilibrium in a laser mode, the laser medium tends to amplify circularly or elliptically polarized waves, whereas the Brewster angle windows tend to preferentially attenuate certain linearly polarized components of the mode. The problem has been treated analytically by Sinclair [40], who shows that when steady state is reached the Brewster angle windows do not actually remove a great amount of energy from the mode unless the resonator losses are very high. This is the reason that Brewster angle windows can be used in commercial CW ion lasers.

In an internal mirror laser there are no polarization determining elements, and modes can exist with arbitrary polarizations. In this case application of even very weak magnetic fields may cause interesting effects such as slight differences in oscillation frequency between modes that have the same TEM_{mnq} configuration but different polarizations. These effects are being studied by a number of investigators.

6. Excitation Mechanisms in Gas Lasers

An understanding of the excitation mechanisms that provide the population inversion in gas lasers is of great interest to physicists who are concerned about the basic processes in lasers. To a person who is using lasers, however, the question is only secondary, for two reasons. First, he can use only such lasers as are known to exist, and he has absolutely no control over the excitation processes that make them work. Second, theories of excitation processes are ex post facto in nature, and so far they have not been particularly useful for predicting new lasers. For these reasons, we give only a brief introduction to the problem of excitation in this book. The reader who is interested in further details should consult review articles in the literature that deal with the question explicitly. Among these are the following: for the neutral atom lasers, the review papers of Bennett [8, 26]; for the ion lasers, the second paper by Bennett [26] and a paper by E. I. Gordon et al. [41]; and for molecular lasers, the work of C. K. N. Patel [14].

An implicit requirement for any steady state inversion is that the lifetime of the lower laser state be shorter than that of the upper state.

Other than that, in both pulsed and CW lasers, it is necessary that the rate of population of the upper state be greater than that of the lower state. There are, in general, three different ways of populating the upper state.

1. Direct excitation by electron collision of atoms in some highly populated lower state, such as the ground state or an intermediate metastable state.
2. Multiple step processes in which an atom must undergo several discrete events before arriving in the desired energy state. These may consist of successive excitations by electron collisions, excitation to a higher energy level followed by cascading through spontaneous emission down to the upper laser state, or any combination of these.
3. Resonant processes whereby an atom excited to a level very close in energy to the desired level is able to raise a lasing atom to the desired upper state, giving up its own energy in the process.

Direct electron excitation is probably the most important populating method in most gas lasers employing moderate discharge currents; this includes neutral atom lasers except the helium-neon laser and most molecular lasers. The notable exception is the helium-neon visible laser, for which excitation of the lasing neon atoms is provided by resonant transfer of energy from excited helium atoms in one of the metastable states. The originally discovered helium-neon infrared laser transitions around 1.15 μ can apparently be populated either by direct collision or by resonant energy transfer from helium metastables. In this case, laser operation has been obtained in pure neon, indicating that direct collision in itself is sufficient to provide population inversion, but the process is apparently also greatly assisted by helium metastables, if they are present. Another situation of "mixed" processes is in the high-power CO_2 laser at 10.6 μ. Here it has been shown that a large population inversion is generated in CO_2 by resonant energy transfer from excited nitrogen atoms, but operation at somewhat lower power can also be obtained without nitrogen, or in a mixture of CO_2 and helium, indicating that other mechanisms must also be present to generate the population inversion. In this particular case, the other mechanism is probably a cascade process from above. Finally, in the CW ion lasers, there has been some debate as to whether excitation to the upper laser level is through a one-step or multiple-step process. There is at present considerable experimental evidence that the excitation process for these lasers depends on the length of time the discharge has been turned on. If the lasers are operated in short pulses, of the order of several microseconds in length, the excitation process is that of a single step from the neutral atom ground state, but after a short period of time this mechanism becomes inadequate to maintain the inversion and multiple-step processes take over. Experimentally, this can be seen by a sharp pulse at the instant of turn-on of the discharge followed by a gap or greatly reduced power output, followed again by a slow rise to eventual CW operation in which multiple-step processes are apparently dominant [42].

Some additional details of these processes as they apply to the particular lasers mentioned are given in the next chapter.

CHAPTER 2

CHARACTERISTICS OF PRACTICAL GAS LASERS

We repeat the classification of lasers according to type, with the most important examples given for each type:

1. Neutral atom lasers (helium-neon laser operating at 6328 Å, 1.15 μ and 3.39 μ).
2. CW ion lasers (ionized argon laser operating on various lines in the blue and green part of the visible spectrum).
3. Molecular lasers (CO_2 mixed with various gases, operating at 10.6μ).
4. Pulsed lasers having a transient response (mercury laser at 6150 Å, nitrogen at 3373 Å).

In this chapter, we discuss the design and operating characteristics of the first three types.

A. RANGES OF THE OPERATING CHARACTERISTICS OF GAS LASERS LISTED ACCORDING TO TYPE

1. Wavelength Range in which Useful Transitions Are Found

Gas laser transitions have been observed over a wavelength range varying from the ultraviolet (2000 Å) to the submillimeter range of the microwave region. However, the more important transitions of the various types of lasers fall within more restricted subranges of this overall range. Figure 16 illustrates the approximate ranges within which most of the observed transitions have been seen for each type of laser, with dashed lines showing regions where some transitions are known but not much study has been made. We see, therefore, that the neutral atom lasers are known primarily by the helium-neon laser with transitions

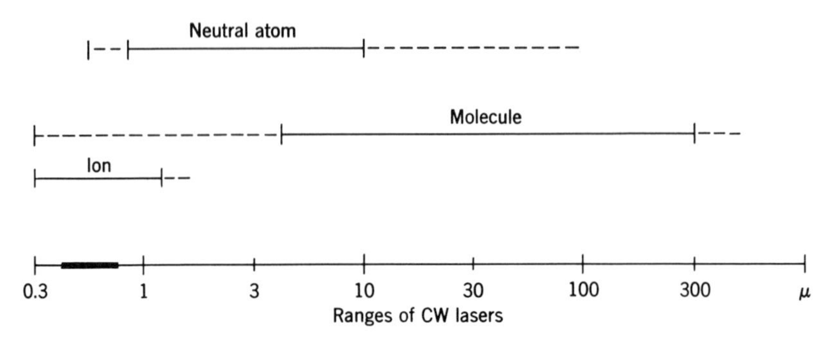

Figure 16. Wavelength ranges of the various lasers, dashed lines showing sparse regions. The visible region is shown by the heavy line.

around 6000 Å, by a few other neutral gas lasers with transitions between 7000 and 10,000 Å, and by a large number between 1 and 10 μ. A study has been made at the Bell Telephone Laboratories of infrared transitions in neutral gas lasers beyond 10 μ, and a number have been observed to wavelengths out to 133 μ [43], but the power outputs are weak and this region has been studied for only a few of the noble gases. Thus, except for the well-known helium-neon visible laser, the neutral gas lasers should be considered primarily as devices that have their highest gain and power in the 1- to 10-μ region.

The molecular lasers are primarily far infrared devices, although, again, transitions are known in the 1- to 3-μ region and, particularly in some pulsed molecular gases, in the visible and untraviolet. The high-power CO_2 laser is at 10.6 μ; its near relatives employing CO, NO, NO_2, and possibly other triatomic species have transitions either in the same wavelength range or around 5 μ. Molecular gas lasers, primarily water vapor, have been studied farther out in the infrared and are known to have strong transitions around 30 μ and also around 120 μ. Other materials, particularly those having a CN bond, appear to make satisfactory radiation sources in the wavelength region between 100 and approximately 800 μ [44], thus overlapping well into the millimeter wave region. In the 100-μ region, in fact, they are probably the only really usable coherent sources now in existence.

The ion lasers are known primarily for the multiplicity of lines in the visible and near ultraviolet. They are very poorly represented in the infrared beyond 1 μ, although some wavelengths from ion lasers have been reported in the region between 1 and 2 μ.

It is possible at this point to present a very general and rough explanation of why the wavelengths of the various gas lasers fall within the ranges in which they are observed. A word of caution is in order, however. It is useful at this point to repeat an anecdote which the author heard as a student and whose source has long been forgotten. A student was given, as a thesis project, the task of determining the distribution in length of fish in the sea. Accordingly, he went fishing with a net whose diameter was 6 ft. and whose openings were approximately 1 in. He caught many fish, measured their length, and plotted the distribution. This distribution had a somewhat bell-like shape peaked at around 1 ft., but trailing off gradually to zero at 1 in. at the short end and 6 ft. at the long end. The moral is obvious. In the case of the search for gas laser transitions, the dimensions of the "net" are determined largely by the availability of suitable optics, including transparent Brewster angle windows, by the availability of sensitive detectors, and, perhaps most of all, by the limits of the experimenter's curiosity. The "explanation" that follows must therefore be considered to be ex post facto, and subject to considerable future modification or refutation.

The explanation is based on a generalized energy level diagram shown in Figure 17, which might apply to any gas laser. In an actual energy

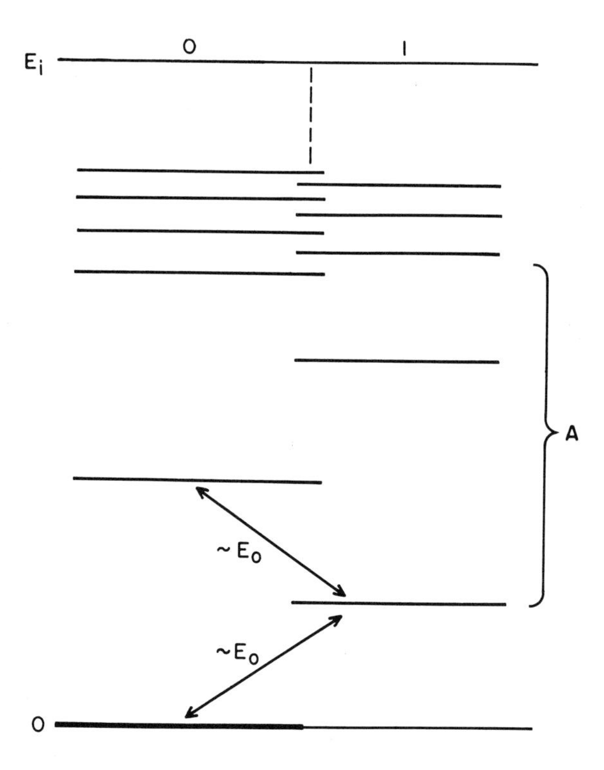

Figure 17. Generalized energy level diagram applicable to any gas laser medium. (From [1].)

level diagram there are many types of state having different angular momentum configurations (*s*, *p*, *d*, etc.), but in all cases the allowed transitions that can give rise to radiation are between states of opposite parity. For simplicity, then, we show only two states: parity zero and parity one. We assume for simplicity that all transitions between states of opposite parity are allowed, and those between states of the same parity are forbidden.

All CW laser transitions and most pulsed laser transitions have been observed in the region marked A in Figure 17. There are only two exceptions which may or may not prove the rule: in a few of the pulsed lasers (very few), the laser transition involves the ground state or a state very close to the ground state. The other exception is the case of the far infrared lasers employing neutral noble gases; in this case the transitions can be considered to be from the region directly above that labeled A in the figure. The fact that laser transitions should occur in this region and only in this region is to be expected. They cannot involve the ground state directly (except in a few exceptional pulsed cases) because ground state populations are so large compared to any excited state populations that can exist in a reasonable discharge that an inversion involving the ground state is very difficult to obtain by means of a discharge alone. (In ruby lasers and some other solid-state lasers they are achieved by optical pumping.) At the other end of the diagram, the states are "hydrogenic" in nature, meaning that an atom starting from a highly excited state can lose its energy rapidly in many steps by staircasing down the energy level diagram, going alternately from a state of one parity to a state of another parity. This is not a situation that will allow for a buildup of population in one state and accumulation of sufficient population to generate an inversion. Such a condition can only apply in the region A where there is a breakdown of hydrogen-atom-level principles and abnormal situations can occur with regard to state lifetime such that an inversion can be built up.

The ranges of observed laser wavelengths in the various gas lasers thus depend on the magnitude of the region A and of the separation E_0 within them. This in turn depends to a large extent on the ionization potential E_i. In the neutral gas lasers, E_i is almost always of the order of 10 eV and all laser transitions are therefore approximately at 1 eV. This agrees quite well with Figure 16. Neon, being the sole neutral gas laser with transitions in the visible, has a somewhat higher ionization potential (15 eV), and in addition the visible laser transitions skip one quantum number in the jump from the upper to the lower state. The transition that does not skip one quantum number, starting from the upper state, has its strongest laser transition at 3.39 μ, and this transition causes considerable difficulty in the design of a helium-neon laser intended to be operated in the visible. Such lasers must often be built so that the 3.39-μ transition cannot initiate laser oscillation, because if it does, it has considerably more gain than the 6328 Å transition and tends to prevent it from lasing. These engineering solutions are discussed later.

In the ion lasers, on the other hand, the ionization potential is generally closer to 20 eV and the energy E_0 to about 2 eV. This corresponds to transitions in the visible and near ultraviolet. The observed visible laser transitions from ion lasers always seem to be primary transitions, involving no jump of orbital quantum numbers nor any pronounced competition effects between various transitions for the available population. Small competition effects are known, but they involve either very weak lines or small changes in power. Such effects can be controlled by the use of prisms to obtain only one wavelength, as in the neutral atom lasers, or they can often be controlled merely by varying the mirror reflectivities in the proper manner as a function of wavelength. If prisms are used in an ion laser, it is almost always because the user wishes to have only one out of several available wavelengths, rather than for reasons of enhancing power.

In the molecular lasers the energy level diagrams of diatomic and triatomic molecules are considerably more complex than the sort of thing that is shown in Figure 17, but the arguments still seem to apply in a rough sort of way. Typical "ionization potentials"—perhaps meaning dissociation in some cases—are generally of the order of 3 or 4 eV, and a corresponding value of E_0, if Figure 17 is taken literally, will be from about 0.1 to 0.3 eV, which puts the radiated energy in the 10-μ region where it is in fact observed. However, the existence of relatively strong transitions from molecular lasers in the 100-μ region could not be explained quite so simply in terms of a diagram of the form of Figure 17, and it must be assumed therefore, that the situation regarding population of levels and relative transition probabilities between closely spaced levels in such systems is much more complicated.

The reader is reminded again of the empirical nature of the discovery of most of the known laser transitions. The existence of an inversion is contrary to nature's normal state of affairs corresponding to a positive and well-defined temperature, thus the existence of any population inversion is a delicate matter depending upon an equilibrium balance between many processes that go on in a discharge, most of which are very poorly understood.

2. Output Power and Gain in Lasers of Laboratory Size

In this section we discuss the typical range of output gains and powers from the various lasers under conditions that may be considered typical for an average user. By "laboratory size" we mean the following. Most users would prefer that a given scientific instrument be of a size convenient to carry by hand, not so small that it must be examined by microscope nor so large that it cannot be conveniently assembled with other laboratory equipment. An overly small laser is hardly conceivable with presently known gas lasers. As for making lasers larger and larger, clearly more power and gain can be obtained in a large laser than in a small one, but limits are reached at which a laser must be considered to be a special installation, part of a building or even a special project akin to a high energy accelerator, rather than a convenient laboratory tool. Such very large lasers have in fact been built. One, at Bell Telephone Laboratories, was 5 m long and generated 1 W of power at 6328 A [45]. Another one, at the same laboratory, was 10 m long and was used for investigation of very weak lines [46] in the near infrared in helium-neon mixtures. However, we shall consider "laboratory size" to be 2 m in length or less. A 2-m long object can be carried by one person through a doorway and assembled on most workbenches. A laser 1 m in length is even more convenient. For certain situations, particularly single-frequency lasers, other considerations dictate that the length must be very short and such structures may be as short as 15 cm in length.

Table 2 lists the output power and gain, when known, for the most important gas laser wavelengths, all of which are available commercially. In the case of the visible helium-neon laser, the table shows only the two extremes of the range, and the reader can use his intuition to fill in the numbers for the many sizes that are available between the two extremes listed here. A more detailed listing of useful laser wavelengths is given in Appendix A.

TABLE 2
Laser Parameters

Laser	Length (cm)	Power Out	Gain Percent Double Pass	Bore Diameter (mm)
He-Ne	15	1 mW	5	1
6328 Å	180	80 mW	40	3
Ar^+				
4880 Å		4880 – 1 W	100	
	60			2-3
5145 Å		5145 – 2 W	40	
CO_2				
10.6 μ	200	100 W	≈ 100	> 10

Although a detailed discussion of the engineering and physical considerations that determine these ranges of output powers is given later, a few words are in order here regarding the power-size relationship in the various lasers. The neutral gas lasers can be built quite small, and generally have low input power requirements, but they also have low gain and low output power. The values given for the helium-neon visible laser are also typical of helium-neon in most of its infrared lines (except the 3.39-μ line) and in most of the other neutral gas laser lines in the near infrared, with the exception of a few particularly strong lines in xenon. The aforementioned infrared lines that have high gain do not have particularly high power; in fact, the power outputs of many neutral gas laser lines correspond to approximately the same number of photons per second, when translated in these units, for lasers of the same size. The high gain of many of the infrared lines, relative to those in the visible and the 1-μ region, arises from the fact that the available population radiates its laser energy into a narrower doppler width, and thus more atoms are provided for a given frequency range.

The molecular lasers are essentially similar to the neutral atom lasers in regard to output power and gain, except for the significantly important CO_2 transition at 10.6 μ. We have listed here this transition in a laser of maximum "laboratory" size, because it is assumed that the highest available power would be desired by most users. Clearly, the laser can be shortened to almost any desired length over a few centimeters, with power output varying approximately as the length. Molecular lasers in the far infrared require very large tube diameters in order to accommodate a stable resonator mode; because of this the enclosed volume is large and the number of emitted photons per second in the laser beam is large, but the energy per photon, $h\nu$, is small and therefore total energies are comparable to those of the weaker lines in the near infrared. Studies have been made of these far infrared lines only recently and not much is known about the gain-versus-power relationships in them.

The ion lasers are capable of very high power in relatively short lengths, and the length of the active discharge medium is usually limited not by size considerations as much as by input power requirements. An input power of 10 kW to a power supply can probably be considered a maximum reasonable number for an average laboratory that does not wish to install special substation equipment for supplying large amounts of power. Smaller sizes are possible physically, but at the present time not very practical, for reasons which are discussed in detail elsewhere in this book.

The pulsed lasers vary widely in their power input and output characteristics, but characteristically they have very high peak gains and output powers, otherwise their average energy output would be low and they would be of no interest. We have listed in Appendix A the pulsed gas lasers for which there appear to be most interest on a laboratory basis and which have properties that are not duplicated by CW lasers. All of the noble gas ion lasers can also be operated as pulsed lasers, often with some enhancement of peak power and gain, and the same may be true of some of the molecular lasers. Helium-neon lasers can be pulsed profitably when operated in the 1-μ region to obtain very high peak output powers, particularly if the bore diameters are large [47]. Pulsing cannot be used on the 6328 Å and other visible transitions, because the upper state in these transitions in helium-neon mixtures is populated by collisional transfer from helium metastables. The accumulation of population in the upper laser state thus depends on first establishing a large population of helium metastables, and this is not a process that is amenable to being enhanced by short bursts of large amounts of electrical energy.

3. Efficiency in Converting Input Electrical Energy into Laser Light

With one significant exception, the conversion of electrical energy put into the discharge to laser energy in the strongest lines of the laser output is approximately 0.03%. The exception, which of course is of great practical importance, is the CO_2 laser at 10.6 μ, for which efficiencies of the order of 20% can be achieved. However, even in the remaining cases where the efficiency is relatively low, there are situations where the number must be interpreted with some caution. Figure 18 shows the general nature of the functional relationship between power input and power output for the various types of laser. These curves hold in general, but the actual figures (watts in versus milliwatts out, for example) must be supplied for any given laser transition. The important fact to be noticed is that the variation of the curve in the case of the CW ion lasers differs from that of all other types of lasers. This particular variation (discussed in detail later in the book) places a great premium on operating the ion lasers at as high a current density as saturation characteristics and power dissipation problems will permit. The peak of the curve in Figure 18 corresponds to the values quoted in Table 2, but it will be noticed that if slightly smaller input powers are provided for the same laser, the efficiency goes down drastically. Thus there is little advantage in attempting to operate the CW ion laser at low powers unless the laser is very small and was specifically designed with low power in mind.

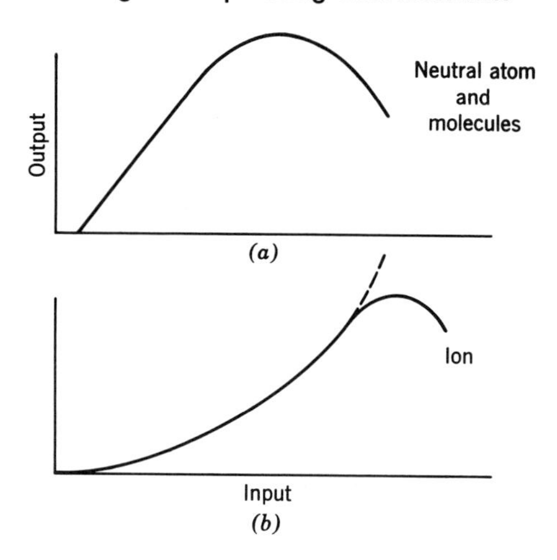

Figure 18. Functional relationships between input power and output power. (*b*) is for a given ion laser with a given filling pressure. If the pressure is optimized at each value of input power, the curve goes monotonically upward as shown by the dashed line.

4. Doppler Width and Coherence of Laser Output

Table 3 lists the Doppler widths (δ and the full width at half-maximum) for some of the more important laser transitions. Also listed are the effective temperatures of the atoms having these Doppler widths, according to (4), and the "coherence length" to be expected of a laser having that Doppler width and operating simultaneously in a large number of temporal modes. The coherence length, briefly, is the maximum difference in distance between two arms of an interferometric system for which one can obtain usable interferometric effects. It is important in such applications as holography, because it determines the maximum depth of field that can be used and of course has meaning for many other interferometric systems. The coherence length is approximately c/δ where δ is the Doppler width as defined in (4).

The effective temperatures of all plasmas except those of the CW ion lasers are approximately at room temperature, perhaps slightly higher than room temperature because of heat dissipation by the discharge. In at least one pulsed ion laser [10, 48] the Doppler width also corresponds to room temperature, perhaps because laser emission occurs before appreciable heating is caused by the discharge. The CW ion lasers, on the other hand, typically have effective temperatures of the order of 3000°K [24], undoubtedly due not only to the actual high temperature of the plasma but also to the fact that the emitting particles are ions and subject to acceleration by electric fields.

TABLE 3
Doppler Widths

Laser	δ (MHz)	Width at Half-Maximum Points (MHz)	Effective Temperature (°K)
He-Ne			
6328 Å	1020	1700	
1.15 μ	552	920	400
3.39 μ	186	310	
Ar+			
Visible	2100	3500	3000
CO_2			
10.6 μ	36	60	300

The remaining variations in observed Doppler widths are due solely to variations of the wavelength and mass of the atom according to (3). It is perhaps unfortunate that the most commonly used laser media have relatively small masses, thus providing larger Doppler widths.

B. PHYSICAL AND ENGINEERING DETAILS OF THE VARIOUS LASERS

We discuss here the characteristics of the helium-neon, argon ion, and CO_2 lasers, particularly those characteristics that determine the specifications of practical laser structures such as those available commercially. Many of the details treated also apply to lasers of the same types employing other materials, although this is not the intent here.

1. Helium-Neon

As a start in understanding the helium-neon laser, recourse must be had to the energy level diagram, as shown in Figure 19. The diagram shows the principal helium energy levels on the left and the neon energy levels on the right. Most important is the close coincidence in energy between the two metastable states of helium and certain excited states of neon. Transfer of energy from relatively large populations of helium metastables to excited neon levels at corresponding energy is effected by collision, and it appears that population of the $3s_2$ state in neon can occur only through transfer from the helium metastables. It apparently

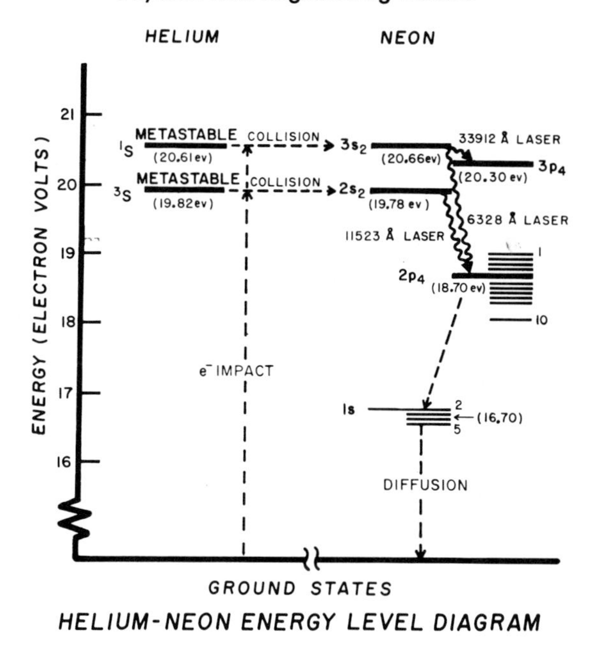

Figure 19. Helium-neon energy level diagram. Only that part of the level schemes of helium and neon that are pertinent to laser operation are shown.

cannot be populated to any great degree by direct electron collision of ground state neon atoms, although the $2s$ levels can be populated by direct electron excitation.

There are thus two states (or groups of states) in neon that have large populations and are available for generating laser action, and there are two sets of lower states in which repopulation by natural means is low enough that an inversion relative to these more highly populated states is produced. Hence, as Figure 19 shows, there are three main types of laser transitions available in neon. Transitions from the upper level $-3s_2-$generate the visible transitions and the very high gain transitions around 3.39 μ, and transitions from the various $2s$ levels generate the 1-μ transitions. Since the 1-μ transitions, starting from the $2s$ levels, can also be excited by direct electron bombardment under some conditions, it has been possible to manufacture lasers operating in this wavelength region containing pure neon, without any helium. On the other hand, lasers employing a mixture of approximately 90% helium, 10% neon can produce all three transitions, and we shall concern ourselves exclusively with this type of laser in the remainder of this discussion.

The pressure of the helium-neon mixture used in these lasers is a slowly varying function of bore diameter, but generally it is in the region of 1 to 3 torr. The physical limitations on the usable pressure region are probably the following. At the lower end, it is very likely simply a matter of requiring a sufficient number of atoms to maintain the desired gain. If all other conditions remain constant in such a discharge, the gain would be directly proportional to pressure and, obviously, below a certain critical pressure, it would not be sufficient to sustain oscillation against losses. A limitation at the high pressure end is probably that as the pressure rises the rate of collisions of excited atoms, and particularly collisions in which they can be de-excited, increases at a more rapid rate than the rate of population of the upper state. In particular, in the case of the $3s_2$ level in the helium-neon laser, where excitation is by means of collision transfer from helium metastables, there are numerous helium atoms remaining in the ground state and reverse transfer of energy can take place from the excited neon atom to ground state helium atoms raising them back to the metastable states. If the pressure in the discharge is raised, the process becomes of increasing importance while the rate of helium metastable production itself is not increased. This, then, changes the steady state equilibrium value of the excited population in the upper neon level. It is significant that higher total pressures can be tolerated in smaller bore helium-neon laser tubes, where it has been shown that the overall rate of excitation is higher and, presumably, helium metastables are also being produced at a faster rate [49].

The existence of two possible laser transitions from the upper $3s_2$ laser level creates serious problems in regard to generation of visible light from such a laser, as has been pointed out. If two lasers of identical construction are compared, one operating in the visible and one at 3.39 μ, and if the relative population differences between upper and lower states were the same in both cases, one would expect the output power of the infrared laser to be down from that of the visible laser by approximately a factor of 5.4, or by the ratio of the energies $h\nu$ per photon at the two wavelengths. Experimentally, this is in fact the approximate ratio that is observed (note the word approximate), and one can assume that the relative population differences are not too different in the two cases. In other words, considering that both transitions start from the same upper level, the two lower levels involved have approximately the same steady state population. In addition, the product $\nu\ \mu^2$ that appears in the gain equation (19) is probably approximately the same in both cases. The difference in gain between the two transitions is then determined primarily by the difference in Doppler width between the two transitions:

$$\frac{\delta_1}{\delta_2}=\frac{\nu_1}{\nu_2}=\frac{3.39}{0.6328}=5.36. \tag{46}$$

This says that the gain per unit length, for short lengths, should be approximately 5.4 times as great in the infrared as in the visible. Experimentally, somewhat higher gain ratios have been observed, but it must be remembered that these ratios have to be expressed in terms of decibels per meter and a factor of 5.4 for lasers of reasonable length can easily raise the gain expressed in decibels to rather large numbers. Experimenters have reported gains at 3.39 μ as high as 50 dB per single pass in lasers whose length was somewhat greater than 1 m. This generates a rather serious problem. Even if the laser mirrors are completely transparent in the infrared, amplification of the spontaneous emission generated at one end of the plasma as the radiation propagates down the laser bore may be sufficient to partially saturate the upper state population that is available for visible laser transitions in various parts of the tube. This effect has been seen experimentally where it has been observed that the upper state population per unit volume is a function of the position along the laser tube under conditions where 3.39 μ radiation is allowed to propagate and acquire sufficient amplification [50].

Because of the high gain that exists at 3.39 μ, it is necessary to take active steps to suppress this particular laser radiation in all helium-neon lasers except the very shortest ones in order to get satisfactory emission in the visible. Three means have been used to accomplish this end. Two of them operate only by suppressing reflection of the 3.39-μ radiation from the end mirror of a resonator, but they do not affect the gain of the transition in the active lasing medium. The third method operates by reducing the gain of the 3.39-μ transition itself.

The first method is that of filling the space between the Brewster window and the end reflector with a material that is completely transparent in the visible but opaque at 3.39 μ. Methane at atmospheric pressure has been found to be quite satisfactory for this purpose [39]. The amount of methane that is needed is not known exactly, but it is necessary that the mixture of methane with any other gas that might be present (e.g., air) must be at pressures not much lower than atmospheric pressure, since methane itself at low pressures may not be completely opaque to all possible 3.39-μ laser transitions. It has been found experimentally that with low methane pressures, the methane absorption consists of sharp bands, some of which coincide with possible transitions in the 3.39-μ region and some of which do not. Since many of these possible transitions have gains that are higher than the gain in the strongest visible transition, the effect may be to suppress one type of 3.39-μ transition while allowing another one to operate, and the effect on the visible will be the same [51]. When the methane is at high pressure, these absorption lines are broadened out to produce a more uniform absorption which covers all spectrum lines in a given wavelength region. The use of methane for suppressing 3.39-μ radiation has been fairly successful in the laboratory but has not been used in commercial equipment because of the difficulty of maintaining a hermetically sealed enclosure filled with methane for long periods of time under field conditions.

The method which is more generally used in commercial equipment for preventing return of 3.39-μ radiation from an end reflector is the use of a prism in the cavity [50,52]. Figure 20 shows two arrangements that have been used successfully. In the first one, the prism is arranged so that laser energy enters and leaves the prism at Brewster's angle, thus avoiding losses due to reflections at the surface. The end mirror of the resonator is placed to reflect the visible radiation back on itself, in the usual manner. However, the 3.39-μ radiation, leaving the laser plasma tube, is deflected by the prism at a 5° smaller angle than the visible radiation, so it is a simple matter to place an absorber in the 3.39-μ beam or arrange that it leave the laser entirely. In the second arrangement, visible light entering the prism is deflected and strikes the rear surface of the prism, which is arranged to be perpendicular to this radiation and thus takes the place of the usual external reflector. The

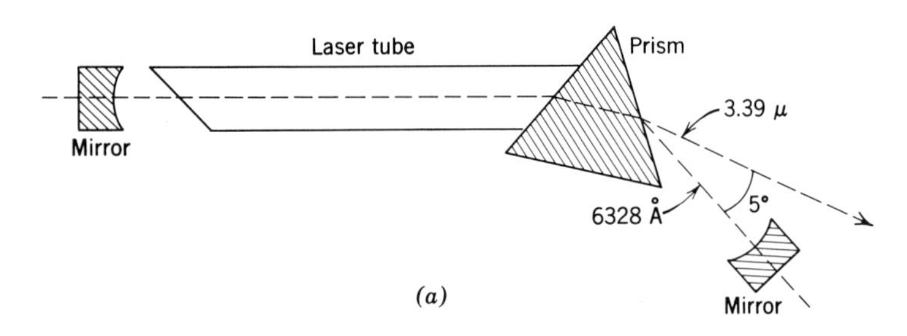

(*a*)

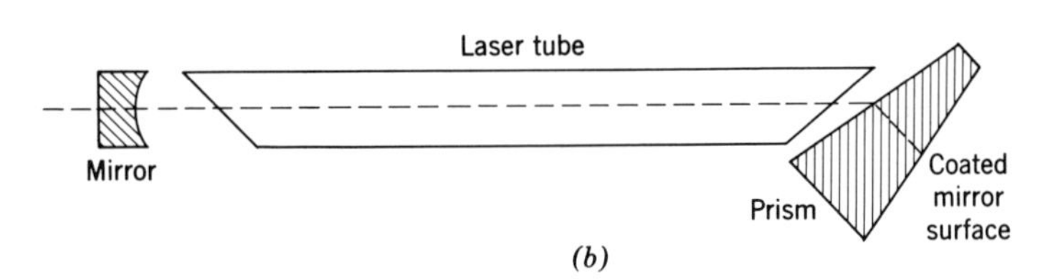

(*b*)

Figure 20. Use of prism in laser resonator to select one of several possible laser transitions.

3.39-μ radiation is deflected by approximately $2\frac{1}{2}°$ from the perpendicular to the rear surface and thus is not reflected back into the laser plasma tube. The selectivity of both methods is approximately the same, and both methods may be used with either flat or curved surfaces for the resonators, although in Figure 20 the rear coated prism is shown with a flat surface.

It may be desirable to curve the rear surface of the prism in order to obtain a curved wavefront. To do this astigmatism must be introduced into the curvature to compensate for corresponding astigmatism produced by refractive effects. This has been analyzed in detail by White [53]. Prisms can be used either at one end of the resonator, or, in the case of very long lasers, they have been used successfully at both ends in order to prevent any more than a single pass gain for the 3.39-μ radiation.

The method of reducing the gain of the 3.39-μ transition is placing magnets along the plasma tube [54]. If an inhomogeneous magnetic field is created which varies widely over different parts of the plasma tube, the Doppler profile of the 3.39-μ transition is "smeared out" by the addition of various amounts of shifted Zeeman components, and since the population per unit frequency range available for inducing gain is thus decreased, the gain is also decreased. When properly handled, this technique does not reduce the gain in the visible appreciably because Zeeman splitting is approximately the same frequency in both transitions, whereas the Doppler width of the visible transition is six times greater. Thus, for the same amount of "smearing," the effective width of the visible laser transition with magnetic fields applied is hardly changed at all. This method has been used with success commercially, and commercial lasers are also available that have a combination of prisms and magnetic fields for the most effective possible suppression of 3.39-μ radiation.

Aside from the special problem of suppressing one particular transition, which we have just discussed, there exists the general problem of the design parameters that must be considered when designing a laser to produce maximum power output or gain within given dimensional specifications. This problem is, of course, common to all gas lasers, but it must be attacked differently for the four different types of laser. The "logic" that is employed in designing a neutral atom laser is shown in Table 4. The upper part of the list shows those entries into the logical chain of reasoning that represent properties of the laser itself, in this case particularly for the neutral atom laser. The last two entries represent limitations based upon resonator theory, which are discussed in more detail in the following chapters.

TABLE 4

Neutral Atom

Gain	$\propto$	1/diameter
Gain	$\propto$	length
Power	$\propto$	gain × volume
$\therefore$		
Power	$\propto$	diameter × length
Diameter	$\propto$	$\sqrt{\text{length}}$
$\therefore$		
Power	$\propto$	$(\text{length})^{3/2}$

A property of the neutral atom laser which is of paramount importance, and which is not shared by the other types of laser so far as is known, is the fact that in all neutral atom lasers that have been studied the gain per unit length appears to be an inverse function of the diameter [8, 26]. The fact that this property is of universal occurrence in the neutral atom lasers suggests that it represents some fundamental property of the discharge, either that the constraint of the discharge by the walls has a profound effect upon the rate of production of excited atoms or that the walls themselves (by virtue of collision processes) are largely responsible for producing the inversion. Experimental evidence seems to indicate that direct effects of the walls upon the inversion are negligible, although not completely nonexistent, and that the constraint of the discharge controls the rate of production of atoms in the excited states, in fact, in both the upper and lower states of the laser transition.

It should be remembered here that the factor under consideration is the maximum available gain that can be obtained from a given laser tube under optimum circumstances, not the relative gains that could be obtained with tubes at different diameters for the same current density, or the same pressure, or the same total current. Indeed, it is found that as the diameter of a neutral atom laser tube is decreased, both the current density and the pressure that are required to obtain optimum gain will rise. This is interpreted as indicating that use of a smaller diameter bore permits a greater rate of production in the excited state to occur before an overall saturation of both the upper and lower laser levels occur, thus giving rise to the higher gain. The controlling property of the gas discharge in both large bore and small bore tubes is the fact that if the current density is raised above the optimum, a change occurs in the relative balance between population of the upper state, population of the lower state, and depletion of the lower state by the discharge processes, so that the total inversion is decreased [49]. The fact that the gain is proportional to the first power of the inverse diameter, and not to some other power, has been verified experimentally for a large number of neutral atom lasers whose bore diameters were larger than the mean free path of the gas.

When gas lasers were first discovered, and the first ones were built for uses other than that of pure discovery, the bore diameters chosen were those that were generally considered "convenient" for glass-blowing purposes, for example, 0.5 to 1.0 cm in diameter. Although such lasers used with spherical mirrors having small radii of curvature will operate in a very large number of transverse modes, the output power in all cases should be approximately proportional to the product (gain × volume). It was soon found, however, that larger bore lasers often did not have as much power output as the smaller bore ones, and it was realized that this was because of the effects of losses at Brewster windows and scattering at mirror surfaces, which tended to reduce the net gain available after the losses have been taken into account. This situation has been observed to be generally true even if mirror radii are chosen so that single transverse mode operation is obtained.

A general result of experience in the laboratory with neutral atom gas lasers having a wide variety of bore diameters and length appears to be the following. For laser transitions having moderate gain (all of the visible ones and most of those in the near infrared) one should use the smallest diameter bore that is compatible with the mode diameter for a given set of resonator mirrors. The power output obtained in this case appears to be greater, even though a smaller volume is available than would be the case for a larger bore, because there is more gain available to overcome losses. In addition, from a purely practical point of view, once a gas laser is out of the laboratory and into the hands of a "customer" (a user who is not particularly concerned with the detailed mechanics of the laser) the internal losses of the laser are likely to be highly variable with time, owing to deposition of dust on Brewster surfaces, and so on. For a laser operated under these conditions, the highest possible gain is absolutely essential, since at low gain the power output is itself a very strong function of the internal losses.

To summarize for the neutral atom lasers: The bore diameter should be kept as small as possible to obtain maximum gain, but it cannot be smaller than the diameter determined by the resonator. Power output can be increased by changing the length of the laser, but increases in length usually entail an increase in bore diameter in order to accommodate the resultant increase in mode diameter of the resonator. The result is the 3/2 power law given at the bottom of Table 4.

2. Argon Ion

The CW argon ion laser is representative of a class of ion lasers that operate CW and have high plasma temperatures [55]. (See Table 3.) These lasers have also been operated on a pulsed basis, and in all cases the strongest lines that have been observed under pulsed conditions are the ones that are also observed in CW operation. Pulsed operation also results in laser emission on many lines that have not been observed to oscillate under CW conditions. In many cases this is due to the fact that higher current densities can be maintained in pulsed operation than in CW operation, but under certain other situations the laser emission may represent a true transient inversion that cannot possibly be maintained under CW operation [56, 57].

The energy level diagram of singly ionized argon is shown in Figure 21. Unlike the helium-neon energy level diagram, there is no obvious reason inherent in the diagram for the particular transitions delineated to lase. It can be assumed that the two sublevels of the $4s(^2P)$ state that constitute the lower states for all of the argon ion laser transitions have very short lifetimes and very high probabilities of emitting energy and dropping to the argon ion ground state. Unlike the neutral atom lasers, where by far the greater part of the entire population is in the ground state, the population at any given time in an ionic ground state is not very large, because of recombination processes that return the ion to levels in the neutral atom energy level scheme. Thus there is no tendency for self absorption to develop a "bottleneck" that would tend to build up a population in the lower laser levels by collisions. In addition, the energy separation between the lower laser levels and the ground state of the ion is very large—about 17 eV—so that even though the plasma is at a high temperature there is no tendency for a Boltzmann population equilibrium to be established between the ionic ground state and the lower laser levels. This same situation appears to exist in krypton and in the other CW ion lasers. In krypton and xenon there are more than two levels that act as lower states for CW laser transitions; however, the general nature of the schemes appear to be similar to that of argon.

In argon the existence of only two common lower states for a large number of visible laser transitions would cause one to think that strong competition effects may exist between the various lines ending on a common level. Such competition effects would be manifested experimentally by the observation that certain lines would be much stronger if observed singly (by means of a resonator containing a prism) than when all lines are observed together. However, it has been found in argon that although such effects exist, they are very minor and use of the prism does

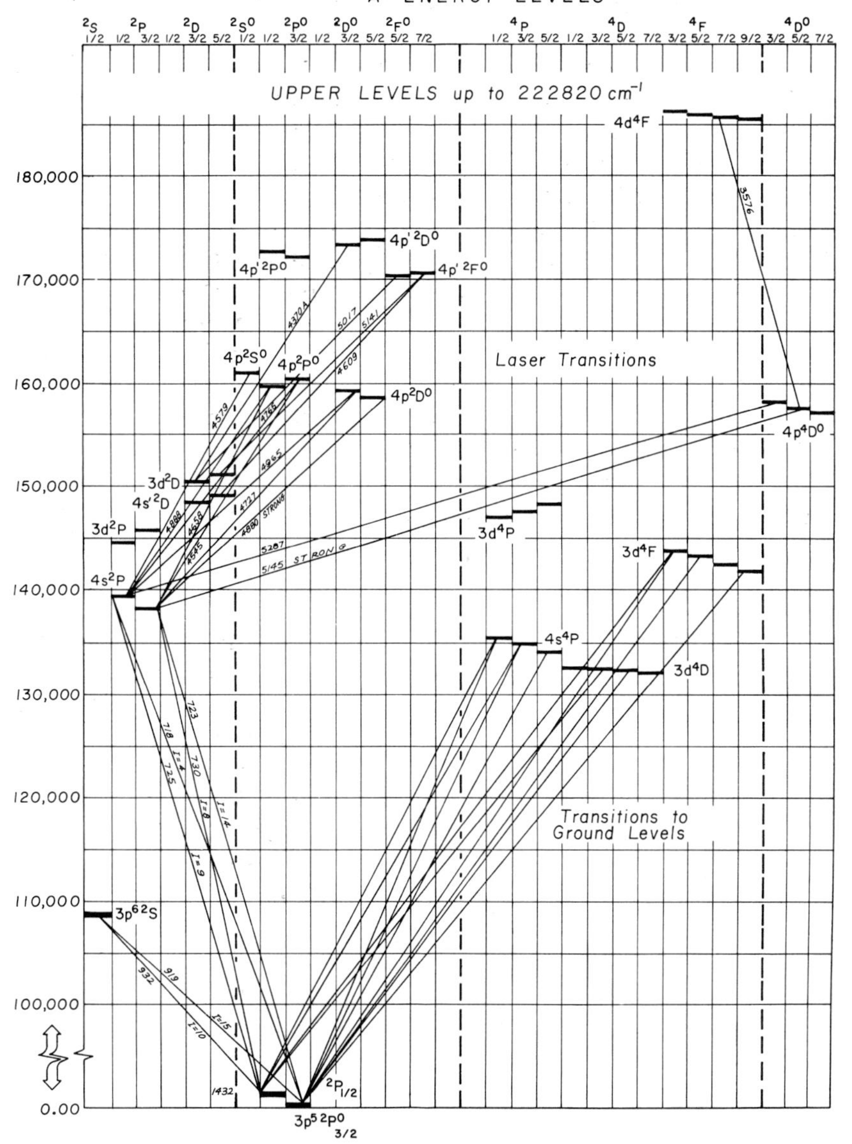

Figure 21. Energy level scheme of Ar^+ showing the laser transitions.

not change the power in any given line by more than about 10%. Similarly, there does not appear to be much competition, even in the case of upper state populations where they are shared by more than one laser transition. For this reason, use of a prism or other dispersing element in a laser resonator is not a particular advantage in CW ion lasers and, if used, only serves the purpose of allowing the user to obtain one wavelength at a time.

A more interesting effect in the case of argon, specifically, is that of relative intensity and gain ratios in the case of the two strong lines, 4880 Å and 5145 Å. The factors that determine the ratio of power output to single pass gain are discussed in Chapter 1, Section D; however, the fact that these can be different for different laser lines in a given laser should be noted here. Most of the visible laser transitions in the CW argon ion laser have approximately the same gain-to-power ratio as 4880 Å, although they are weaker than that line and generally have less gain. The 5145 Å line has only about one-fourth the gain of the 4880 Å line, but has approximately twice the power output when the gain is sufficient to overcome internal losses. This effect is largely due to differences in the atomic constants that determine the saturation parameter in the steady state equations, owing to the fact that the 5145 Å upper state comes from a different family of levels than do those of most of the other transitions.

This effect is even more extreme in the one remaining line in the argon-ion laser at 5287 Å. In a high power laser this line has so little gain that it is very easily wiped out by imperfections in either the windows or the mirrors. However, when proper care is taken that losses at this wavelength are reduced to a minimum, the power output obtainable in this line is approximately one-fourth that of the 5145 Å line. It should be noted that a cascade effect exists between the upper levels of the 5145 Å and 5287 Å lines and levels that are farther up in the energy level scheme, and pulsed laser transitions have been observed in which the ^4P level that is the upper state of the 5145 Å line is the lower level of an ultraviolet transition. However, this effect has not been observed in CW lasers, and some cascading must occur in the case of all of the upper states of all of the argon ion laser transitions.

The logical reasoning involved in designing ion lasers is shown in Table 5. The relationships of gain and power between themselves and relative to the laser bore geometry are quite different from those of the other lasers. First, the gain is very strongly dependent on current density because high current densities are essential for maintaining a large population of atoms in ionic states (in low current discharges the atoms are primarily in excited states of the neutral atom). In addition it

TABLE 5

Ion

Gain	$\propto$	(current density)^2
Gain	$\propto$	length
Power	$\propto$	(gain)^2 × volume
Diameter	$\propto$	$\sqrt{\text{length}}$
Power	$\propto$	(current)^4 × (length)^2

appears likely that the excitation process which generates the population inversion in the CW ion lasers requires more than one collision step in order to reach the final upper state of the laser transition, and this in itself would account for the dependence of gain upon current density squared. The dependence of the power on the square of the gain rather than on the first power, as is usual with most other lasers, is a peculiar property of ion lasers and is due to the fact that they generally behave as inhomogeneously broadened lasers [see (45)] even when a large number of modes are present. To explain briefly why this is so, it arises because the natural linewidth in ion lasers is very broad, generally much broader than the separation between temporal modes. This causes intense competition between adjacent temporal modes so that one mode suppresses its near neighbors, and the observed mode spacing is some multiple of $c/2L$ rather than $c/2L$ itself. Thus, instead of having a uniformly saturated Doppler profile as required for homogeneous broadening, the profile has holes burned into it at approximately uniform spacing and apparently with unsaturated "shoulders" between holes. This is a condition of inhomogeneous broadening. As the power density within the laser is raised, eventually a condition corresponding to homogeneous broadening must be reached, at which the laser will be operating at optimum efficiency. In an argon laser, this condition can probably be reached at the 1-W output level if losses internal to the laser are kept to a minimum.

Unlike the neutral atom lasers, the gain per unit length in ion lasers does not depend directly on the bore diameter, but since the gain is such a high power of current density it is usually desirable to keep the bore as narrow as resonator geometry will permit in order to maintain the highest possible current density for a given total current. For this reason the relationship between diameter and resonator length is the same as that suggested for the neutral atom lasers—the smallest possible diameter permitted by the length. The last two lines in Table 5 merely combine the reasoning of the previous lines algebraically. The high-power dependence of total power output on current was observed directly in the earliest experiments on CW ion lasers [22], and the length-squared dependence has also been observed experimentally in a resonator in which sections of identical current but variable length were operated. The current dependence shown in Table 5 does not, of course, continue indefinitely for higher and higher currents; after a certain critical value of current is reached, current saturation sets in and less power is obtained instead of more. The value of the saturation current varies for the different gases, being in general at the highest current densities for light gases such as argon and chlorine and lowest for the heavy gases such as xenon.

The lowest current density at which laser action can take place varies widely among the various laser transitions and is also a function of variable parameters such as mirror reflectivity. An example of the relative threshold current densities for a particular setup is given in [58].

One of the important factors in the design of CW ion lasers is the effect of a longitudinal magnetic field (lines of force of the magnetic field parallel to the axis of the bore) [59]. The effect of a magnetic field in increasing the power output of a particular argon ion laser is shown graphically in Figure 22. Similar effects are observed in ion lasers employing other gases, although in the heavier gases the optimum value of magnetic field is at somewhat lower values and often somewhat more critical. The primary effect of applying a magnetic field is to increase the electron density in the plasma for a given value of applied current, and this is accomplished by constraining the electrons to move in orbits or helices around the field lines and thus parallel to the axis of the bore, thereby preventing loss of electrons to the walls. A similar effect can be achieved by employing cyclotron resonance at microwave frequencies [60], but in that case the magnetic field has to be transverse to the bore in order to generate the cyclotron orbits that keep electrons away from the walls.

In Figure 22 the loss of effectiveness of the magnetic field at very high values of field is undoubtedly due partially to plasma conditions but is also due to effects of Zeeman splitting of the laser emission line, with consequent broadening and loss of gain. This is particularly to be expected in the case of a laser tube with Brewster windows, where the Brewster windows define normal modes of output polarization that do not agree with the normal modes that are generated by the Zeeman splitting. It will be noted also in Figure 22 that at certain values of the field the power output of the laser was actually increased when a Brewster window was inserted into the cavity (no other changes were made in the laser for this experiment). The reason for this effect is not

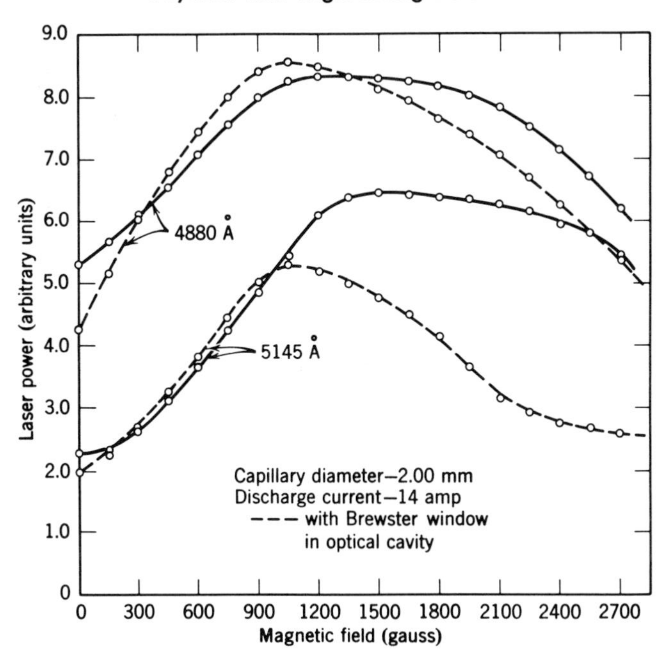

Figure 22. Output power of a perpendicular window laser at 4880 Å and 5145 Å as a function of magnetic field, with and without a quartz slide oriented at Brewster's angle in the optical resonator. (Reproduced by permission from [59].)

well understood but probably has to do with complex mode coupling effects.

3. CO_2

The pertinent part of the energy level scheme for the high power CO_2 laser is shown in Figure 23. Early experiments with lasers containing CO_2 demonstrated the existence of the 10.6-μ laser line in CO_2 alone, without the existence of other gases; however, the power output was quite low. The first indication that high power could be obtained from this group of transitions was given by Patel, who obtained his high power output from CO_2 mixed with nitrogen [61]. He postulated the need for the close coincidence in energy between levels in nitrogen and levels in CO_2 as shown in Figure 23 [14]. More recently, high power output has been obtained from CO_2 mixed with helium, and although a helium-nitrogen-CO_2 mixture appears to be close to optimum, considerably higher power can be obtained from CO_2 mixed with helium than can be obtained from CO_2 alone [62]. For this reason, it is not clear that the close energy coincidence between nitrogen and CO_2 is absolutely essen-

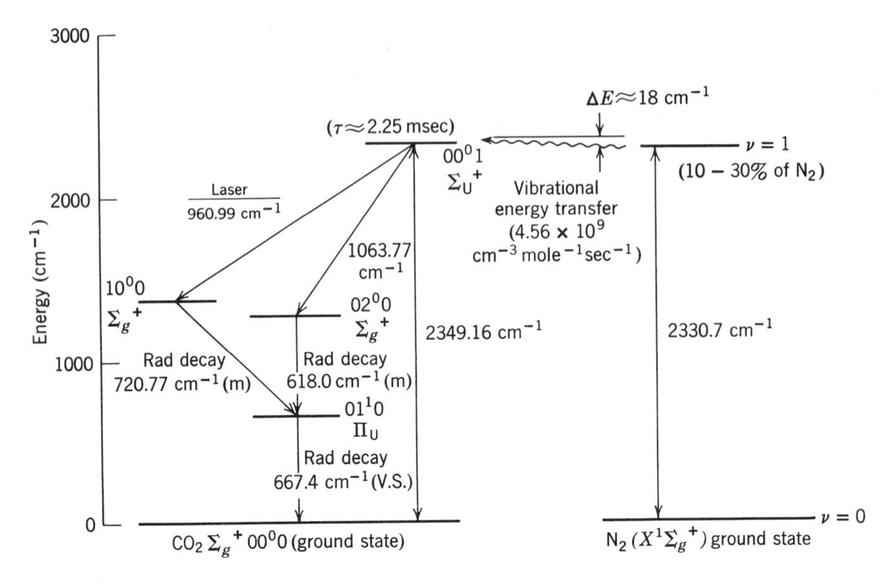

Figure 23. Partial energy level diagram of low-lying vibrational levels of N_2 and CO_2. (Reproduced by permission from [14].)

tial for obtaining high power, and it is likely that a major part of the population process is due to cascade effects from levels in CO_2 above those shown in the figure. This fact is particularly important because it has been found that mixtures of CO_2 and nitrogen are not particularly stable in a discharge, tending to form CO and possibly other detrimental substances, and the output of a CO_2-nitrogen or CO_2-nitrogen-helium laser drops to a low value within a very short time unless the gas is continually replenished by flowing from a storage reservoir.

Mixtures of CO_2 and helium alone appear to be more stable against detrimental effects of the discharge. In the CO_2 laser the maintenance of depleted population in the lower state is particularly critical, because the energy difference between the lower state and the ground state is not very large and at temperatures of the order of room temperature or above the Boltzmann population equilibrium maintains a relatively large part of the total population in the lower laser level. Thus it is particularly important to keep the temperature of the gas in the discharge as low as possible. It has been theorized that one of the beneficial effects of helium in a CO_2-helium mixture is that it conducts heat rapidly from the excited gas to the walls. For the same reason, the wall temperature is particularly important, and the power output of any CO_2 laser can be increased by factors of 2 to 4 by cooling the walls from room temperature to dry ice temperature or below. Hot walls are particularly detrimental, and in all cases high-power CO_2 lasers have to be cooled by means such as flowing water if refrigeration methods cannot be used.

The importance of exerting direct control over the population differences in the CO_2 laser has been emphasized by recent measurements on the lifetimes of the laser states [63]. The addition of nitrogen to pure CO_2 increases the lifetimes of both the upper and lower states; however, the effect of the resonant energy transfer in the upper state results in a net increase in population difference. Helium, on the other hand, decreases the lifetime of the lower state without affecting the upper state. Thus a combination of He and N_2 added to the CO_2 is particularly effective. Other gases have been tried as additives by various investigators from time to time, occasionally with beneficial results, but the effects are always marginal.

The logic employed in designing a CO_2 laser is shown in Table 6. It is actually the simplest of the three cases since power output depends only on length and hardly at all on diameter [64]. Thus to obtain high-power CO_2 output one simply makes the device as long as possible. The only limitation on diameter is that it must be larger than the minimum permitted by diffraction of a mode at 10.6 μ. For a resonator 2 m long, this minimum diameter should be approximately 1 cm. In practice, larger diameters are easy to build and sometimes advisable. Actually, the power falls off very slowly as a function of increasing diameter; but the fall-off is not sufficient to warrant special consideration for design purposes.

TABLE 6

CO_2

Gain	$\propto$	length
Power	$\propto$	gain or length
Diameter	$\gtrapprox$	$\sqrt{\text{length}}$
Power	$\propto$	length

4. Summary

We summarize briefly the conclusions reached here in the logical diagrams for the three specific lasers that have been discussed in detail. For helium-neon, a long length is advisable for high power, but it is generally advantageous to keep the bore diameter to the minimum permitted by the optics in order to provide for high gain. Special attention must be given to suppressing the 3.39-μ laser transition, and in structures longer than 2 m this problem may be difficult even if steps are taken to suppress the transition at the ends of the resonator. Structures less than 2 m long can be built with relative ease. In the argon ion laser the key factor is high current density, with limitations of the power supply capacity a secondary problem. This dictates structures approximately 1 m in length or shorter, with bore diameters as small as can be permitted by the optics. In the CO_2 laser, the only restriction on size is that the length should be sufficiently great that the gain can well overcome all losses, and it is suggested that the length be as great as possible.

In Table 2 a summary of these conclusions in terms of the specific lasers mentioned is given, with the bore diameters included. These bore diameters are those of commercially available lasers and agree quite well with the design parameters suggested in this section.

CHAPTER 3

MODES AND RESONATORS

The concept of a laser mode and its description in terms of spatial and temporal variations have already been developed to some extent. In this chapter we develop the mathematical theory further and discuss the problems involved in maintaining spatial and temporal mode purity.

A. SPATIAL MODES

1. Types of Spatial Mode. Importance of Lowest-Order Mode

A basic property of the "modes" of any resonator is that they represent energy storage within the resonator. A laser resonator is relatively long in one direction—the z direction—compared to the other directions x and y, and therefore the mode can be considered as a spatial xy configuration propagated back and forth in the z direction. If the energy is stored for any reasonable length of time, the equivalent propagation path is very long and any initial energy distribution will rapidly be modified by diffraction effects. In other words, any distribution considered a near-field distribution rapidly becomes converted into its far-field distribution. Since the mode must represent a static configuration, however, we are left with the fundamental requirement that the *far-field distribution must have the same form as the near-field distribution.* This condition is also *sufficient* to define the spatial modes.

Mathematically it is simplest to consider the problem first in one dimension, x, with the propagation still in the z direction. If the electric field vector in the near field is given within an equiphase surface of the wave front by $U(x)$, then the corresponding function in the far field $V(x)$ is given by

$$V(x)=a_1\int_{-\infty}^{\infty} U(x')\exp(-ia_2xx')\,dx' \tag{47}$$

where the constants a_1 and a_2 depend on geometrical details and are otherwise irrelevant here. The important point is that the functions U and V are *Fourier transforms* of each other. Thus a spatial mode in this one-dimensional situation would be any function (describing the electric field vector) that is its own Fourier transform. The functions that satisfy this condition are

$$\phi_n(x)=H_n\left(\frac{\sqrt{2}\,x}{w}\right)e^{-x^2/w^2} \tag{48}$$

where n is an integer and the H_n are known as Hermite polynomials. It can be shown [65, 66] that the functions $\phi_n(x)$ form a complete orthogonal set, which means that any reasonably well-behaved function of x can be written as a linear superposition of the functions $\phi_n(x)$. The constant w is a unit of length that determines the actual physical size of the functions $\phi_n(x)$. An arbitrary function can be decomposed into components of the complete set of functions $\phi_n(x)$ for any value of w.

We now consider the physically more meaningful case of two-dimensional field functions. In cartesian coordinates x, y the relation converting the near-field distribution $U(x,y)$ into the far-field distribution $V(x,y)$ is

$$V(x,y)=a_1\int_{\infty}^{\infty} U(x',y')\exp\left[-ia_2(xx'+yy')\right]dx'\,dy'. \tag{49}$$

In polar coordinates r, θ the relationship is

$$V(r,\theta)=a_1\int_{\infty}^{\infty} U(r',\theta')\exp\left[-ia_2rr'\cos(\theta-\theta')\right]\cdot r\,dr\,d\theta. \tag{50}$$

The functions that are their own transforms under (49) are

$$\Phi_{mn}(x,y)=H_m\left(\frac{\sqrt{2}\,x}{w}\right)H_n\left(\frac{\sqrt{2}\,y}{w}\right)\exp\left(-\frac{x^2+y^2}{w^2}\right); \tag{51}$$

in other words they are simply products of the one-dimensional functions ϕ_n given in (48). The functions that are their own transforms under (50) are

$$\Phi_{pl}(r,\theta)=\left(\frac{\sqrt{2}\,r}{w}\right)^l L_p^l\left(\frac{2r^2}{w^2}\right)\exp\left(-\frac{r^2}{w^2}\right)\cos l\theta, \tag{52}$$

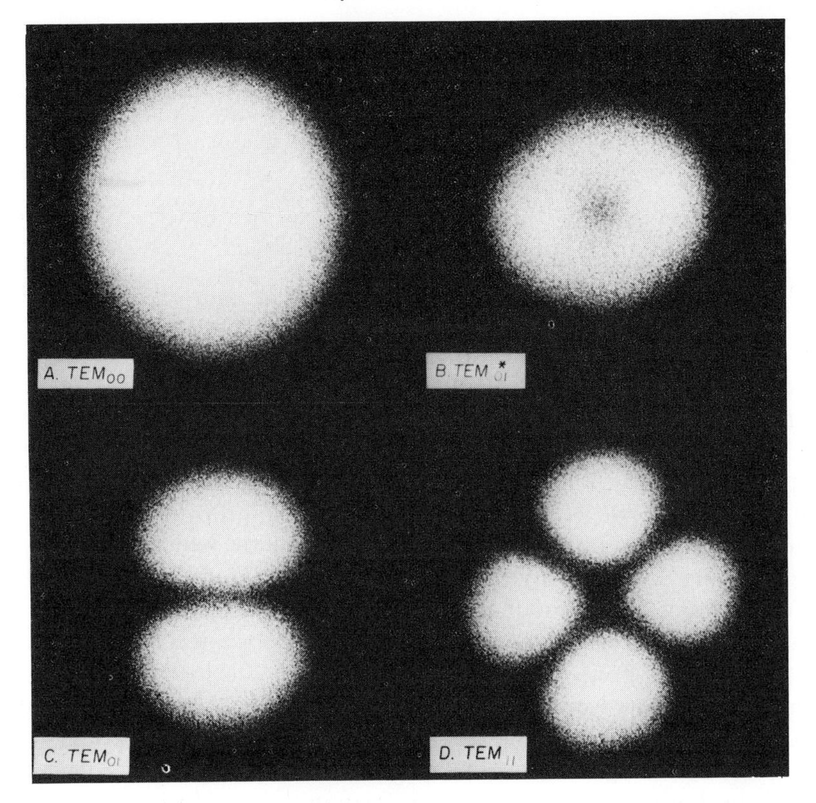

Figure 24. Intensity distributions of four of the lowest order spatial modes.

where the functions $L_p{}^l$ are known as generalized Laguerre polynomials. This representation is not very useful unless the radial mode number l is very small—0 or 1—indicating a high degree of axial symmetry. For $l = 0$, (52) becomes a function of r alone and the functions $L_p{}^0$ are the ordinary Laguerre polynomials. It can be shown that the functions defined by (52) form a complete orthogonal set [66, 67].

Figure 24 shows the mode patterns produced by the lowest-order modes. The mode numbers given are the numbers m, n of (51). It should be remembered that the observed intensity is the absolute square of the function ϕ_{mn}, not the function ϕ_{mn} itself. The reader who is interested in seeing the patterns produced by higher-order modes will find examples of them in review papers such as that of Kogelnik and Li [68] and in most of the textbooks on lasers listed in the Bibliography.

The lowest-order mode has the mode numbers 0,0 in either the x,y or r,θ coordinate representation and is a simple Gaussian described as follows:

Electric field vector:

$$E(r)=E_0\, e^{-r^2/w^2}, \tag{53}$$

intensity:

$$I(r)=I_0\, e^{-2r^2/w^2}. \tag{54}$$

Since w defines the mean radius of the mode, it is commonly referred to in this connection as the "spot size." The lowest-order mode is particularly important in laser applications because it does not contain any phase shifts in the electric field across any cross section of the laser output beam. It has two additional advantages that make it particularly useful. First, it is quite simple to describe mathematically. This, together with the fact that the configuration remains unchanged in going to the far field, gives rise to simple formulations for the propagation of the laser output beam inside and outside the resonator, which are discussed in following sections. Another advantage of the lowest-order mode is that, for a given mean diameter of the output beam, this mode has the smallest angular divergence as the beam is propagated through space. It can also be shown by an extended mathematical development that energy propagated in the form of this mode can be focused down to a spot that is as small as the smallest physically permissible according to the laws of electromagnetic wave propagation.

Of the higher-order modes, one frequently seen in laser outputs is the "doughnut" mode indicated in Figure 24 as TEM^*_{01}. This requires special explanation since it has not yet been adequately described. It consists of the two modes TEM_{01} and TEM_{10} (rectangular coordinates) arranged 90° out of phase optically, so that the two lobes of the mode rotate around the z axis at the optical frequency of, say, 3×10^{14} Hz for visible light. What is seen, of course, is the mean square average intensity, which is given by

$$I(r)=2I_0 r^2\, e^{-2r^2/w^2}. \tag{55}$$

This mode can be focused to a spot not much larger than that of the lowest-order mode, but since there is a phase reversal in the wave front at all times, the mode is less useful for many applications.

Having defined the nature of spatial modes, we shall now proceed to calculations of the parameter w, which describes the physical size of the mode.

2. The Equations of Boyd, Gordon, and Kogelnik

The relationship of w to the physical parameters of the resonator was first derived for confocal resonators by Boyd and Gordon [16] and then generalized to other resonators by Boyd and Kogelnik [17]. Consider a resonator defined by mirrors 1 and 2 having a separation d and radii of curvature b_1 and b_2, respectively. b_1 and b_2 are defined as positive if the mirrors are concave facing the interior of the laser (Figure 25). The spot sizes on the surfaces of the mirrors are w_1 and w_2, respectively.

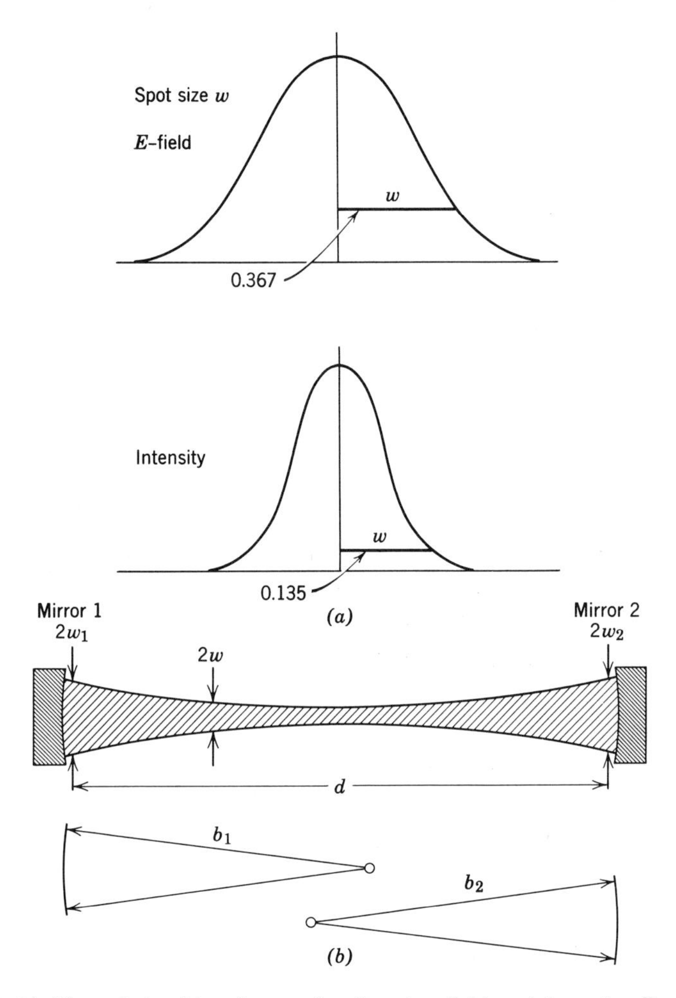

Figure 25. (*a*) The relationship of w to the Gaussian field and intensity distributions. (*b*) The definitions of the parameters w, b, and d as they apply to a laser resonator.

The computations of Boyd and Gordon [16] and Boyd and Kogelnik [17] can then be summarized in the following single formula, which has universal validity for resonators employing plane or spherically curved mirrors:

$$w_1{}^4 = \left(\frac{\lambda}{\pi}\right)^2 \frac{b_1{}^2 d}{b_1 + b_2 - d} \times \frac{b_2 - d}{b_1 - d}. \tag{56}$$

This equation gives the spot size for w_1; to obtain w_2 it is merely necessary to interchange indices 1 and 2 wherever they appear in the equation. A special case of some interest is that for which $b_1 = b_2 = d$, commonly known as "the confocal resonator," although the word "confocal" as such has a slightly broader meaning [17]. For this particular situation, $w_1 = w_2 = w$, and w is given by the following simple relationship:

$$w = \left(\frac{b\lambda}{\pi}\right)^{\frac{1}{2}}. \tag{57}$$

At the center of a confocal resonator the wave front is a plane wave whose value of w is given by

$$w' = \left(\frac{b\lambda}{2\pi}\right)^{\frac{1}{2}}. \tag{58}$$

An important consequence of the work of Boyd and Kogelnik is the fact that any given value of w' determines the propagation characteristics of the wave, as well as the "equivalent confocal resonator" whose mirror radius is b. This may be used to determine the value of w and the angular divergence of the beam at any point in space, either inside or outside the resonator. The relationships for developing propagation properties of the laser beam are derived in Chapter 4.

3. Practical Resonator Configurations and Stability

A wide variety of combinations of mirror radii can be used with any gas laser discharge tube to provide modes having desired values of diameter and angular spread. In addition to the confocal resonator, other combinations in common use are depicted in Figure 26. Almost any type of stable resonator that may be devised will fall within one of the classes shown in Figure 26; however, those most commonly used are the hemispherical and the long radius configurations. The others merely duplicate the essential properties of the three important types of resonator—confocal, hemispherical, and long radius—but may have certain spe-

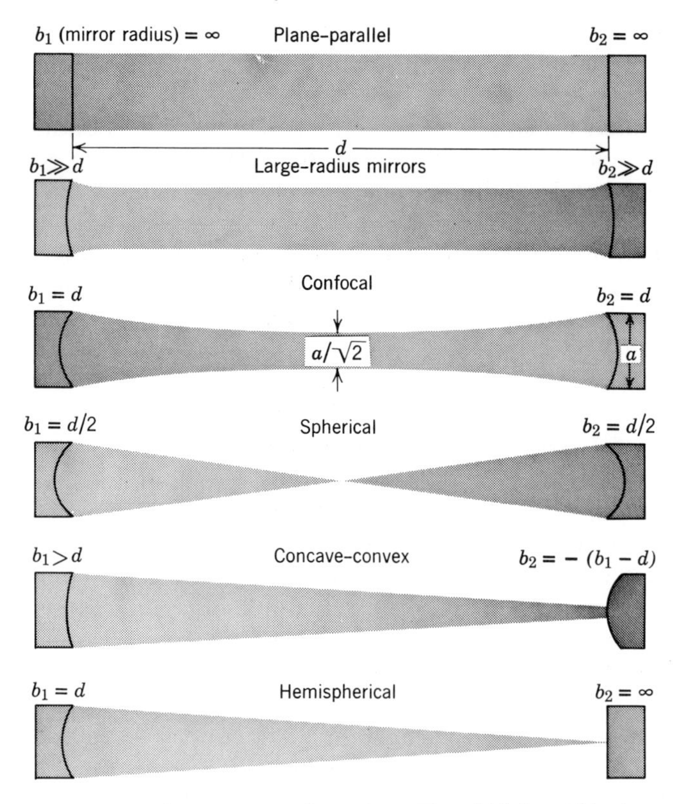

Figure 26. Commonly used resonator configurations. The width $2w$ within the resonator is shown by the shaded portions of the diagram.

cialized uses. For example, configurations close to the full spherical resonator are characterized by the mode narrowing down to a very tiny waist for a short distance in the center of the resonator configuration. This configuration may be useful for very short, very narrow bore laser tubes in which it is desirable to place the resonator mirrors some distance away from the discharge tube itself.

The plane-parallel resonator may be considered a limiting case of large-radius mirrors, but unlike the "stable" resonators involving at least one spherical mirror, the plane-parallel resonator is just barely on the edge of stability. For this resonator the effective mode diameter is controlled not by mirror radius of curvature, since there is no radius of curvature, but entirely by the radius of the mirror itself. To the extent that stability is achieved at all, it is achieved because diffraction losses at the edges of the mirror give rise to slight phase shifts in the wave front at the mirror surface (approximately one-sixth of a wavelength between the center and the edge of the mirror), which cause the resonator mode to be quasi-spherical in the vicinity of the mirrors. The mode also has an approximately Gaussian shape, with a spot size w corresponding approximately to one-fourth the radius of the mirror. Since the geometry of the existing mode in the resonator is controlled entirely by the diffraction losses around the edges, these diffraction losses are obviously very high, much higher than those usually encountered in stable resonators employing spherical mirrors. This makes the plane-parallel resonator difficult to use with lasers having low gain, no matter how high the reflectivity of the mirror surfaces themselves.

The plane-parallel resonator is also extremely sensitive to angular misadjustments, since very slight deviations from parallelism will increase the diffraction loss enormously in the direction in which the mode is deflected after one passage through the resonator and one reflection off the misaligned mirror. When all these factors are taken into account, the plane-parallel resonator has little to offer in the way of usefulness for the average user of a gas laser. It is primarily of historical interest; however, it has one property that is potentially useful and that can be duplicated only by the confocal resonator when operated in exact confocal conditions. Resonant frequencies for odd order spatial modes (those for which $m + n =$ odd number) lie exactly halfway between the resonant frequencies of the even order modes. This property may sometimes be useful when operating with multispatial-mode lasers. For all other types of resonator this condition is violated either slightly or flagrantly.

The long-radius mirror configuration is widely used at the present time as a means of obtaining a laser mode whose characteristics are much like that of the confocal resonator but whose spot size is adjusted to match the bore diameter. The relationship between spot size w and bore diameter is usually chosen so that possible diffraction losses are small compared to mirror transmissions or other losses. The large-radius mirror configuration also has the advantage that the diameter of the mode is more constant through the length of the discharge tube than is true for most of the other resonators, and it thus makes more efficient use of the gas volume. The disadvantage of this mode is, from the standpoint of the user, primarily that the sensitivity of the mode to misadjustments in angle is somewhat greater than that of the confocal or hemispherical resonator. From the standpoint of the manufacturer, the disadvantage is that the mode diameter cannot be controlled easily merely by small changes in length of the resonator, and thus a careful selection of the mirror radii has to be made in order to obtain optimum mode diameter.

The hemispherical resonator is commonly used when it is desired to get lowest-order mode operation from a laser plasma tube whose diameter is large compared to w' of a confocal resonator having the same length. The diameter of the mode at the spherical mirror (or at any point within the resonator) is determined entirely by the separation between mirrors. If the flat mirror is placed exactly at the center of curvature of the sphere, then the mode diameter is theoretically infinite and, in fact, the actual diameter is controlled by diffraction losses as in the plane-parallel resonator. For a resonator length d slightly less than the spherical mirror radius b_1, the following approximate relationships hold, obtained by specializing (56):

$$w_1^4 = \left(\frac{\lambda}{\pi}\right)^2 \frac{b_1^2 d}{b_1 - d}, \tag{59}$$

$$w_2^4 = \left(\frac{\lambda}{\pi}\right)^2 d(b_1 - d), \tag{60}$$

where w_1 is the spot size at the spherical mirror and w_2 is at the flat. From these expressions it can be seen that the mode diameter at either mirror can be controlled quite easily by small adjustments of the resonator length d, and hemispherical resonators typically operate with $(b_1\text{-}d)$ of the order of a few millimeters in order to obtain a mode diameter comparable to plasma tube diameters (approximately 1 cm) in the vicinity of the spherical mirror. The hemispherical resonator is also a relatively simple one to keep in alignment, particularly because once the flat mirror is placed exactly perpendicular to the laser plasma tube all further adjustments are made with the spherical mirror.

So far we have been concerned mainly with resonators whose mirrors were of approximately equal radius of curvature, or for which one mirror was a flat. The general situation must take into account the possibility that mirrors may have unequal, or even negative, radii or curvature. Obviously, not all combinations of mirror radii and separations are stable; many of them originate diverging wave fronts that effectively focus all energy away from the mirror at the opposite end and result in very high "diffraction losses." The conditions under which resonators are stable can be summarized rapidly by reference to (56), noting that the expression for w^4 has meaning only if it is positive. Thus a simple test for stable resonator conditions is to put all parameters into (56) and see whether it is positive; if so, it represents a stable resonator; if the expression is negative, the resonator is unstable. For the specific case of two unequal radii of curvature b_1 and b_2, both positive (corresponding to concave mirrors), Figure 27 shows the stable and unstable regions.

A somewhat different representation of the regions of stable and unstable resonators is shown in Figure 28, originally in the article of Boyd and Kogelnik [17]. The various resonators are delineated in their respective regions, and resonators of particular interest are indicated. It is interesting to note that the exact confocal resonator, so important throughout this book, lies at the exact center of the diagram in a region which is at a border line between stability and instability. Although this is theoretically a problem in regard to whether or not a confocal resonator should be stable, the problem does not appear in practice for two reasons. First, mirror radii are often not exactly identical, nor is it particularly easy to space them at exactly their radii of curvatures. Second, the diagram takes no account of losses other than diffraction losses. Such losses tend to "smear out" the edges of the boundary between stability and instability and in fact provide usually a generous margin of stability in a region around the exact confocal point.

A much more detailed discussion of resonator modes than can be provided here has been given in a review paper by Kogelnik and Li [68].

4. Aligning a Laser Resonator

Aligning the mirrors of a laser so that they point at each other, and so that their mutual axis is accurately down the center of the bore, may or may not be a difficult operation. If the bore diameter is relatively large (if it is large compared to the mode diameter of the smallest mode that can be passed through the length of the bore), then the alignment is simple. An autocollimator may be used, or in some cases it is possible to align the resonator by eye. It can often be done by merely removing one mirror from its mount, looking for a reflection of light from the far mirror, and repeating the process at the other end. However, if the bore diameter is small, or if mirrors cannot be removed from their mounts and replaced reproducibly, then it may be necessary to use a

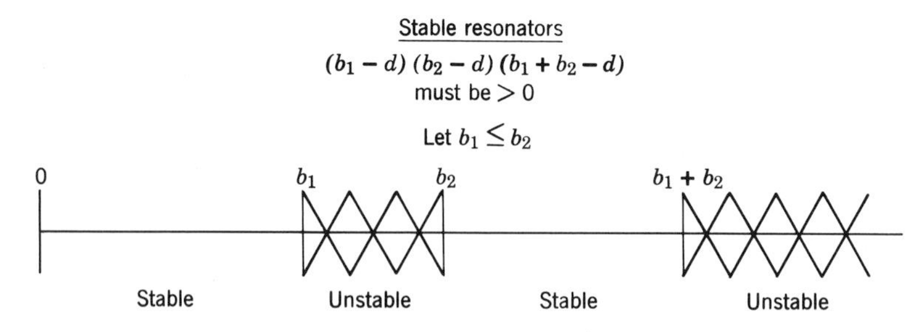

Figure 27. Stability diagram for resonators consisting of two concave mirrors.

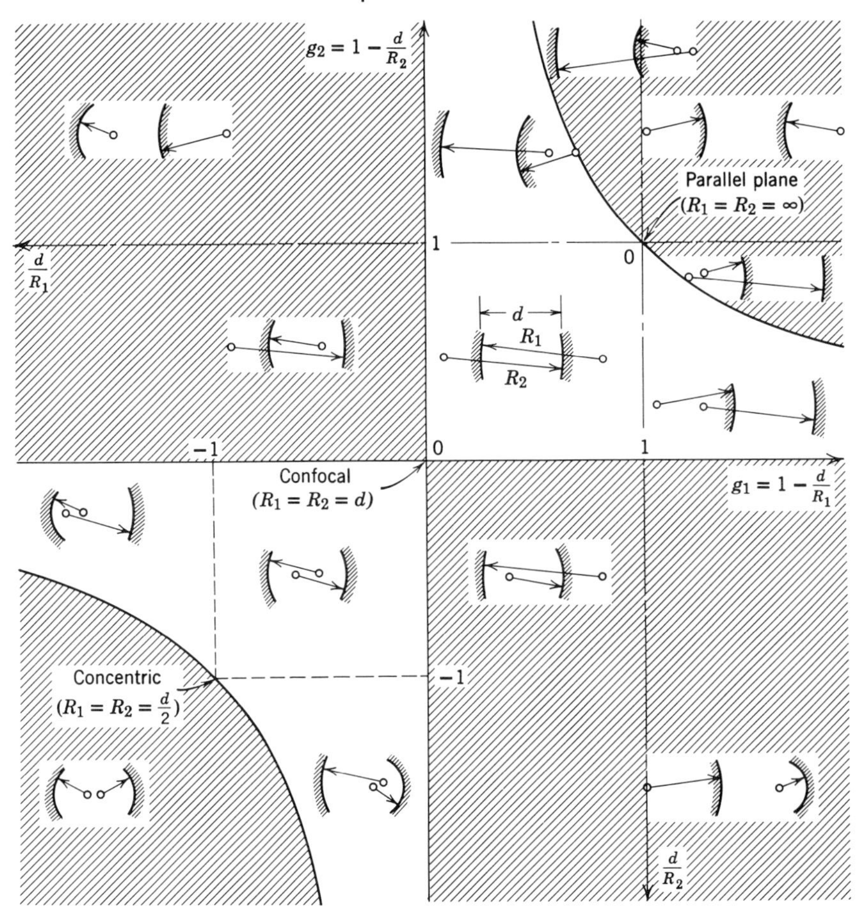

Figure 28. Stability diagram. Unstable resonator systems lie in shaded regions. This figure, reproduced by permission from [68], is an adaptation of the original figure in [17].

second laser to align the one in question. Since a second laser may not always be available, and since most experimenters have pet methods of aligning their laser resonators, we shall not go into the details of the mechanics of the alignment. We discuss here the theory—some of the theoretical problems that are involved and the general methods of attack of optimizing a resonator that already has laser operation to some extent.

The first thing to remember is that the axis of the lowest-order mode (in fact, of all modes) must lie along the line joining the centers of curvature of the two mirrors. If one mirror is a flat, the mode axis lies

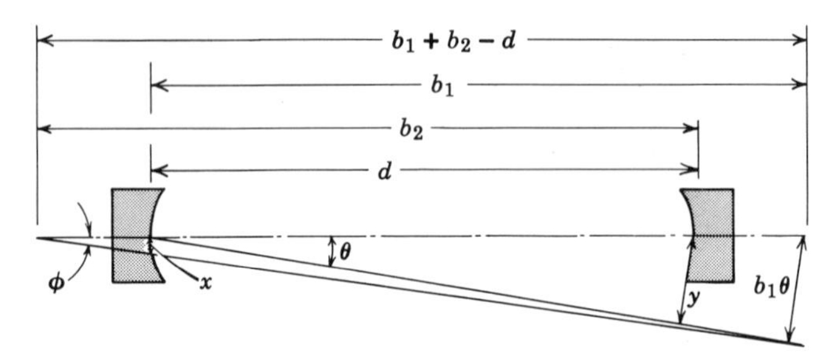

Figure 29. Mirror alignment parameters.

along the line joining the center of curvature of the spherical mirror that is perpendicular to the flat. Figure 29 shows the geometrical relationships for the case of two mirrors whose spacing is slightly less than the radii of curvature. We assume that the mirror with radius b_1 is turned through an angle θ, and that small angle approximations are valid. Correspondingly, the line joining the centers of curvature is turned through an angle ϕ. The following relationships then hold, to small-angle approximations:

$$b_1 \theta = (b_1 + b_2 - d)\,\phi, \tag{61}$$

$$x = \frac{b_1 (b_2 - d)\,\theta}{b_1 + b_2 - d}, \tag{62}$$

$$y = \frac{b_1 b_2 \theta}{b_1 + b_2 - d}. \tag{63}$$

Here, x and y are displacements of the mode center on the mirror being moved and on the opposite mirror, respectively. The sensitivity of the mirror alignment adjustment is the ratio $d\phi/d\theta$, which can be deduced from (61) after substituting the numerical parameters corresponding to the resonator. It is worth discussing some general points, particularly as they apply to commercially available lasers.

In a confocal resonator, the centers of curvature lie on or close to the mirrors themselves, and we have $x = 0$, $y = b\theta$, as if the mode were anchored to the mirror being displaced. For this reason, the confocal resonator is not only the least sensitive to misadjustments but is also the easiest to put back into correct adjustment. It is merely necessary to adjust each mirror independently so that the power output is an optimum; this corresponds to the mode being correctly centered on the opposite mirror. No further adjustments are then necessary, since adjustment of the mode on one mirror (by manipulating the opposite mirror) does not disturb adjustments that have been made previously. The hemispherical resonator is almost as easy to adjust, except that motion of the spherical mirror does not "swing" the mode spot on the other mirror but instead displaces the entire mode while keeping the axis parallel. This can perhaps best be seen by imagining a surface tangent to the spherical mirror that is parallel to the flat; it is then seen that displacing the sphere is equivalent to "rolling" the sphere on the flat. For this reason, care must be taken in aligning a hemispherical resonator to be sure that the mode is accurately parallel to the axis of the laser bore before final alignment of the spherical mirror. The final touch-up on the sphere merely places the mode accurately in the center of the bore.

Alignment of a long-radius resonator or of an almost-spherical resonator is a considerably more difficult process than that of the two resonators just described. In these cases, displacement of one mirror not only moves the mode on the opposite mirror but also on the mirror being displaced by an amount and direction that depend in detail on the resonator geometry. For this reason adjustment of the resonator for optimum output power by manipulating the two mirrors alternately is not necessarily a convergent process, or if it converges, it may result in the mode being not in the center of the laser tube bore. Such resonators are best aligned by a method called "walking." The two mirrors are adjusted alternately, in a manner that would work best for confocal resonators, until a local optimum output power is achieved. Then one mirror is displaced from the local optimum until the power is reduced by an appreciable amount, perhaps to one-half of the initial value. This is performed on one axis, for example, the horizontal axis. The other mirror is then adjusted along the same axis until an optimum is again achieved. If this optimum is greater than the previous optimum, the process is repeated with the first mirror displaced again, in the same direction, and again it is observed whether the next optimum is greater than the one before it. In this way one may plot successive values of optimum output power that will show a curve having a generally broad maximum. This maximum may be considerably greater than the local optimum value that one started with in the first place. The mirrors should be set at this particular optimum value and the process repeated on the other axis. When this process has been performed on both axes, the laser mode is accurately in the center of the bore and the true maximum output power has been achieved. Any further adjustment of the mirrors without performing this walking process may result in a reduction of the output power.

Generally speaking, the difficulty of performing the walking process, and of maintaining alignment generally, increases with increasing radius for long-radius mirrors. A particular point should be noted concerning the almost-spherical resonator. In such a resonator the centers of curvature are very close to each other, and the ratio of this separation to the mirror separation d may be very small. Thus a very small motion of one mirror can result in a very large angular misadjustment of the resonator mode. As one approaches the spherical configuration, the ratio (mirror separation/separation of centers) very rapidly approaches infinity and the resonator becomes unmanageable. This makes the almost-spherical resonator perhaps the most difficult of all to keep in alignment.

5. Conditions for Operating in a Single Spatial Mode

In most laser resonators control of spatial modes is obtained by controlling the losses around the edges of the mode. For many practical purposes, a good approximation to the losses incurred in any given mode can be calculated by computing the integral of the intensity in the part of the mode that lies outside the clear region of the resonator, the mode shape being computed as if there were no limit to the diameter of the laser tube. To give an example, the lowest-order mode has a variation in radius only, with intensity given by

$$I(r) = I_0\, e^{-2r^2/w^2} \tag{54}$$

If the radius of the resonator (limiting aperture of the resonator, mirror radius, or whatever) is aw, then the relative diffraction loss is given approximately by

$$\Delta A = \frac{I_0 \int_a^\infty I(r) r\, dr}{I_0 \int_0^\infty I(r) r\, dr} = e^{-2a^2}. \tag{64}$$

Computations of this type made by Fox and Li [15] and Boyd and Gordon [16], show that the lowest-order mode always has a smaller diffraction loss than any of the higher-order modes. For this reason, if the diameter of the resonator is sufficiently constricted, only the lowest-order mode can operate, all others being prohibited by excessive diffraction loss.

Experimentally, the situation appears to be somewhat more complicated. It has been observed as a general rule that if the diameter of the resonator is sufficiently large, higher-order modes will always operate in preference to lower-order modes if diffraction losses in these modes are not excessive. In particular the doughnut mode shown in Figure 24 appears to be a serious competitor for the lowest-order mode in most lasers and is particularly difficult to eliminate because its diffraction losses are only slightly greater than that of the lowest-order mode. Mixtures of the two modes are often observed in laser outputs and are not only difficult to eliminate but also difficult to observe by eye because both modes have axial symmetry and not very much variation in intensity with radius. A calculation that correctly takes into account the competition of the lowest-order mode and the doughnut mode for the laser energy must take into account not only the diffraction losses in each mode but also the saturation characteristics of the atoms that radiate into the mode and the relative volumes of the two modes. This can be done in the following way. Assume homogeneous broadening and a limiting aperture within the resonator whose diameter is $a \times w$. A steady-state condition is reached in which there is a certain total energy within the resonator and each atom is saturated to the extent determined by the field intensity in its neighborhood. Mathematically, we replace (36) by

$$(A+T)\int_0^{a/2} rI(r)dr = G\int_0^{a/2} \frac{rI(r)dr}{1+I(r)}, \tag{65}$$

where $I(r)$ is either (54) for the Gaussian or (55) for the doughnut, as appropriate.

Equation 65 can be solved numerically. The results depend in detail upon the mirror transmission, other losses, the gain, and saturation characteristics, but most of the results can be described generally by curves of the form shown in Figure 30. This figure shows, for the lowest-order mode and the doughnut mode, the relative power output that would be obtained in the two modes on the assumption that, "by magic," the other mode could be prevented from operating. Assume now that the mode that has the higher power output is capable of suppressing the other mode from operating, which is valid because the mode that generates the stronger average electric field saturates the medium and reduces the effective gain for any other mode that may attempt to go into oscillation. From the figure it is thus apparent that the lowest-order mode will predominate for laser bore *diameters* up to about $3.5w$, beyond

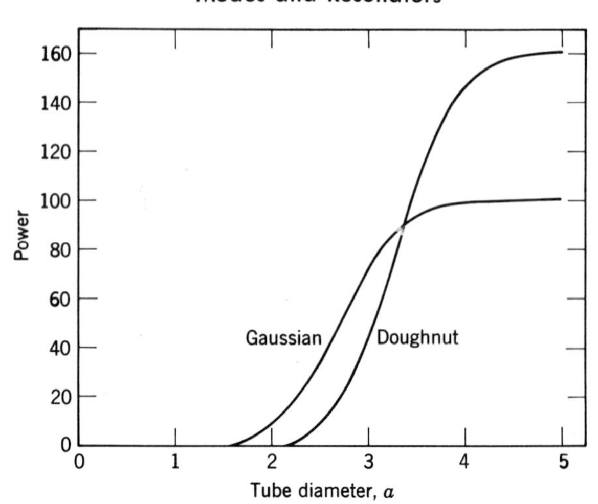

Figure 30. Relative power output as a function of tube diameter a expressed in units of w.

which the lowest-order mode is suppressed and the doughnut, or even higher-order modes, will take over. This result (the curves of Figure 30 were obtained by calculation) agrees quite well with what is observed experimentally. It has been found to be generally true that the bore diameter of any laser whose mode operation is limited by diffraction must have a diameter between $3.5w$ and $4w$. To give a concrete example, a 1-m confocal resonator operating at 6328 Å has a spot size w at the ends of the bore of approximately 0.5 mm. Thus an effective bore diameter for obtaining single spatial mode operation for this laser is 2 mm, which agrees very well with what has been found to be usable in practice. The diffraction loss in the lowest-order mode is then approximately e^{-8} or 0.0004, which is considerably less than the unavoidable losses from either mirror transmission or scattering losses within the laser itself.

A computation even more sophisticated than that of (65) has been performed by Fox and Li [69], who consider the propagation and possible modification of the mode as it traverses the saturable active medium. They find, at large apertures, that a wave front initially in the form of a single mode described by Φ_{pl} of (52) may turn into a mixture of functions Φ_{pl} having the same l but different p. Since "mode" implies a static configuration, the calculation raises a serious question as to what the true "modes" are, or even if they exist, in some types of high-gain laser. Fortunately, the simple mode theory developed in this book seems adequate to explain most of the effects that are observed experimentally.

If the aperture of the laser bore or of the resonator mirrors is other than circular, the conditions for mode selection may be somewhat different. In particular, CO_2 lasers have been operated with mirrors in which the useful output was obtained by boring a hole through the center of the mirror, the outside of the mirror being covered with some sort of infrared transmitting window. In this situation, the diffraction loss is primarily the same as the transmission and is in the center of the resonator, that is, along the axis of the mode. Clearly, this type of configuration tends to suppress the lowest-order mode and encourage the operation in higher-order modes such as the doughnut. Since the doughnut mode can be focused to a spot which is not much larger than that of the lowest-order mode, such operation may be entirely satisfactory for many uses of the CO_2 laser, however, the user should be aware of the fact that the laser may not be operating in the lowest-order mode.

B. TEMPORAL MODES

1. Rules Governing Frequency Spacing

Generally, for any given spatial mode, the frequency separation between the temporal modes that constitute this spatial mode is given by $c/2L$, as explained previously. This relationship will be used in the following discussion, after a few additional comments. This relationship ($c/2L$) is expected to be true only if the refractive index of the medium within the laser is completely independent of frequency. However, the gain characteristics of the laser medium are accompanied by anomalous dispersion, which provides a variable refractive index around the vicinity of the laser emission frequencies. This anomalous dispersion gives rise to an effect equivalent to that of slightly variable vacuum resonator lengths, the exact length being different for slightly different frequencies, and thus tends to shift the frequency separation slightly from the $c/2L$ relationship.

Although this property is not important for the discussion here, it is important when one attempts to perform heterodyne experiments between modes, since the presence of more than two modes not quite evenly spaced results not in one single beat frequency, but in (N^2-1) frequencies where N is the number of operating modes. These frequencies are often separated by only a few kiloherz, and are a particular nuisance if the length of the laser is variable, as is generally the case, since the weaving in and out of these very low beat frequencies is equivalent to a low frequency noise of relatively high amplitude on the output power of the laser.

The second comment has to do with the location of the temporal modes as a function of spatial mode. For m, n, and q defined as the number of nodes of the electric field in the spatial and temporal modes, a formula derived by Boyd and Kogelnik [17] can be rewritten to give the frequencies of the stable modes:

$$\nu(m,n,q)=\frac{c'}{2d}\left\{q+\frac{1+m+n}{\pi}\cos^{-1}\left[\left(1-\frac{d}{b_1}\right)\left(1-\frac{d}{b_2}\right)\right]^{\frac{1}{2}}\right\}, \tag{66}$$

where b_1, b_2 and d are defined as in (56) and c' is the mean speed of light in the laser structure at frequency ν. In a plane-parallel resonator all frequencies are coincident and spaced by $c'/2d$. In a resonator that is exactly confocal or hemispherical the frequencies of the even-order modes alternate with those of the odd-order modes and are exactly halfway between. For other resonators the relationships are considerably more complicated.

In the remainder of this discussion we shall assume that we are working with only one spatial mode (lowest-order mode).

2. The Conditions for Operating in a Single Temporal Mode

Briefly stated, and as a rough approximation, a laser will operate in a single temporal mode if only one such mode lies in that part of the gain profile that is capable of sustaining oscillation (Figure 31). The laser is assumed to have a Gaussian Doppler profile, and the spacing between possible stable temporal modes is given by $c/2L$, as is usual. The Doppler profile is drawn from zero gain, but stable oscillation is possible only in those regions where the gain is greater than the total losses including mirror transmission. Two possible situations which can give rise to single temporal mode operation are shown in the figure. One case is that of a long laser, in which the mode spacings are relatively close together but the gain at the center of the Doppler curve is just barely greater than the losses, so that the region over which stable oscillation is possible is smaller than twice the separation between modes. The other possibility is the short laser, in which the mode separation is quite large, in fact larger than the mean width of the Doppler profile. In this case stable oscillation in a single mode can be achieved even though the gain of the laser is considerably above threshold.

These two possible situations have led to two different philosophies concerning the design of lasers intended to operate in a single temporal mode. Many laboratory experiments intended to test the ultimate frequency stability of a laser have employed the long laser operated just above threshold. For laboratory use this system offers certain advantages. The gain of a long laser tube is intrinsically higher than that of a short tube, but it can be reduced to a value just above threshold by reducing and controlling the input power. However, if one does not wish to go into considerable engineering effort in designing the laser, one knows at least that a value of input power can be found such that only a single mode will be operative at a time. A larger structure is, in general, easier to work with than a smaller structure, and since the input power must usually be greatly reduced anyway, a large structure is capable of dissipating heat and operating without severe rises in temperature. Then, too, if a laser is operated at a gain level such that only a single spatial mode can operate at a time, and if the resonator length changes, the operating frequency of this mode can never be greater than $c/4L$ from the actual Doppler line center. For longer lasers this maximum possible deviation obviously becomes smaller.

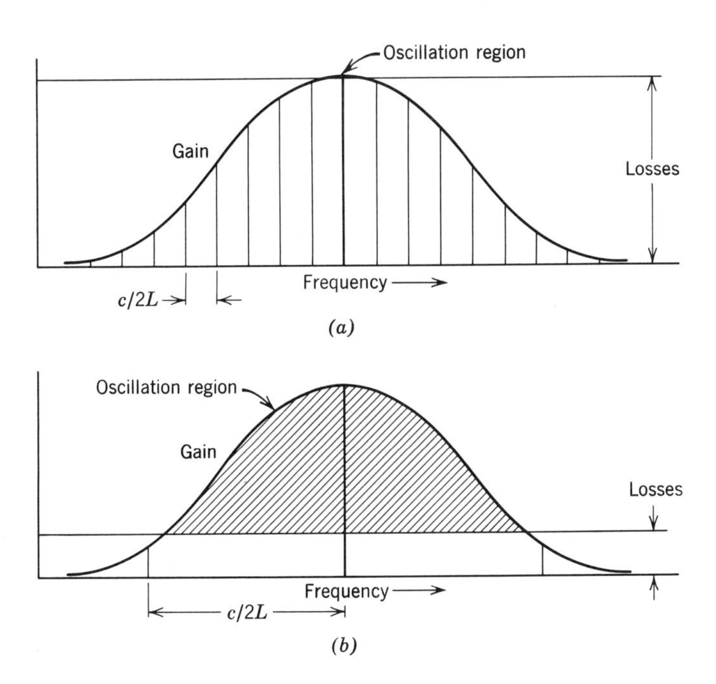

Figure 31. Operation in single temporal mode. (*a*) Long resonator. (*b*) Short resonator.

The disadvantages of the long laser arise partly from the fact that the gain must be controlled to high precision (to prevent the laser from either stopping or operating in more than one mode) and also in the intrinsic length stability problem of the resonator. The ultimate stability of the laser resonator structure, assuming a perfect temperature equilibrium and absolutely stable value of the temperature, is determined by the cumulative effects of the Brownian motions of the atoms and molecules that make up the resonator structure—by the Brownian motion of one mirror against the other. Surprisingly enough, this limit has actually been reached in certain very carefully prepared laboratory experiments on the 1.15-μ helium-neon laser [70].

In order to achieve this limit, two lasers of essentially identical construction were placed in the cellar of an old manor house located near Cape Cod, Massachusetts, well away from highways, railroads, and any other possible causes of seismic disturbances. The room was thermally stabilized and isolated; the experiments were controlled by remote control from adjacent rooms, and the lasers were left unattended for months on end in order to reach complete temperature equilibrium while operating at the desired power level. The relative stability between the two lasers was then found to be approximately 1 Hz over periods of several seconds, with very slow drifts of the order of 1 kHz over periods of minutes or more. This represents just about the amount of drift that is to be expected on the basis of thermal motions. This value and, more important, all thermal expansion effects are proportional to the length of the resonator, so that, on an absolute basis, a short resonator will have smaller changes per given thermal or mechanical disturbance than a long laser. The only major disadvantage of the short laser is that its gain per pass is less, requiring more careful engineering of the design if it is to have reliable operation. However, since commercial laser manufacturers are engaged in exactly this type of engineering, all lasers that have been marketed commercially and that are intended expressly for single-frequency operation have been designed as short lasers, and in all cases such lasers are considerably shorter than lasers of the same design in which single-frequency operation is not necessarily looked for.

For a short laser designed for single-frequency operation, the length of the resonator must be dictated by the Doppler width, as well as by available gain and other considerations. In order to insure reliable operation at one frequency under a moderately wide range of variations of laser gain and internal resonator losses, the resonator length is usually chosen so that, if the desired mode is at the center of the Gaussian, the two adjacent modes will lie at about the $1/e$ points of the Gaussian.

For a helium-neon laser operating at 6328 Å, this dictates a resonator length of approximately 10 cm, and for an argon ion laser the length would have to be about 3 cm. At either of these lengths, approximately 5% gain double pass is available for the laser output lines in question, resulting in a reasonable if not ample margin of available gain over probable resonator losses. Nevertheless the output power of such single frequency short lasers is subject to considerable variation if the internal losses are varied. Part of this is due to the fact that such lasers, operating with a small hole burned in a much broader Doppler profile, are quite inhomogeneously broadened and subject to the effect of the squared term in (45).

3. The Lamb Equations and the Power Dip

When a laser is operating under single-frequency conditions, the relationship between gain and output is not necessarily as simple as that given in (45) for the purely inhomogeneously broadened line. It has been pointed out earlier that the Doppler distribution of velocities interacting with a standing wave in the laser resonator may give rise to not one but two holes burned in the Doppler profile, symmetrically located with respect to Doppler center. The equation for the inhomogeneously broadened line, (45), on the other hand, was derived on the assumption that there was a single hole burned in the Doppler profile; one started with the equation for homogeneous broadening and performed an integration over all resonant frequencies. Now, if we attempt the same sort of integration with two sometimes unclearly separated holes, we are faced not only with a double integration but also with the fact that each moving atom is, in effect, seeing two separate frequencies and interacting with them in a very complicated way. Thus an attempt to set up the problem in terms of Bloch's equations leads to a very difficult integration problem, which may not even have a solution in closed form.

A very useful approximate solution, however, has been obtained by Lamb [7]. Starting from fundamental quantum-mechanical considerations, he derived a general theory of laser operation with two or more simultaneous frequencies. The details of the theory are beyond the scope of this book. We are concerned here with the following result, which gives the power output P as a function of frequency deviation from Doppler center (certain constants of proportionality omitted):

$$P = \frac{Ge^{-\nu^2/\delta^2} - (A+T)}{1 + \dfrac{\gamma_{ab}/\gamma'_{ab}}{1 + (2\pi\nu/\gamma'_{ab})^2}}. \tag{67}$$

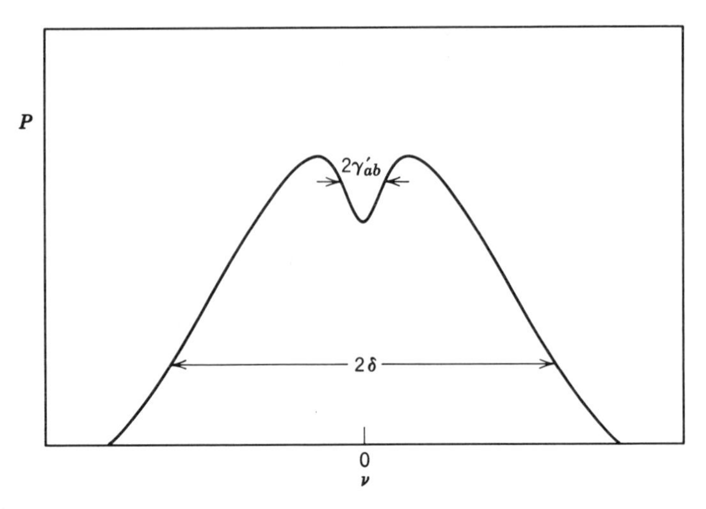

Figure 32. Gaussian curve with power dip.

Here $\nu = 0$ is Doppler center, A, G, and T have their usual meanings, and γ_{ab} and γ'_{ab} are the hard-collision and soft-collision line widths as described previously. It should be pointed out that Lamb's original derivation assumed only one natural line width; the modification to include hard- and soft-collision widths was made later by Szöke and Javan [34] and by Gyorffy and Lamb [71].

A graphic illustration of the output power P given by (67), under the assumption that $G \gg (A + T)$ and $\gamma_{ab} = \gamma'_{ab} \ll \delta$, is shown in Figure 32. The interesting result of this equation is the dip in the power output at $\nu = 0$, superimposed upon the expected Doppler curve. Under the extreme conditions employed in the figure, it is as if a "hole" corresponding to the shape of the natural resonance were subtracted from the Doppler curve. When the conditions are less extreme [$\gamma_{ab} \approx \delta$ or $G \approx (A + T)$] the dip is less prominent, and when the gain is very close to threshold there may be no dip at all, only a slight leveling off of the curve at $\nu = 0$. A full experimental verification of this equation was given by Szöke and Javan [34] in experiments with a highly stabilized laser operating at 1.15 μ. Figure 33 is taken from their paper. Figure 33*a* corresponds to pure ^{20}Ne isotope, giving rise to a symmetrical Doppler curve, and the dip shows up as the gain is increased, as predicted by the formula. If natural neon is used, the Doppler curve corresponding to the unsaturated gain is unsymmetrical because of the presence of small amounts of the isotope ^{22}Ne, and in that case the power dip is also unsymmetrical and poorly defined. Although a full explanation of the power dip, or "Lamb dip" as it is often called, is buried in the quantum mechanics that lead to the derivation of (67), it is possible to develop a qualitative argument from simple considerations such as the hole-burning concept.

Consider the power dip to be caused by the coalescence of two symmetrically located holes as they approach each other at line center. From this argument we would conclude that the width of the power dip is the width of each hole, which is generally true. We would also conclude that at line center one should expect approximately half the power obtainable outside the power dip. Equation 67 actually predicts this to be true

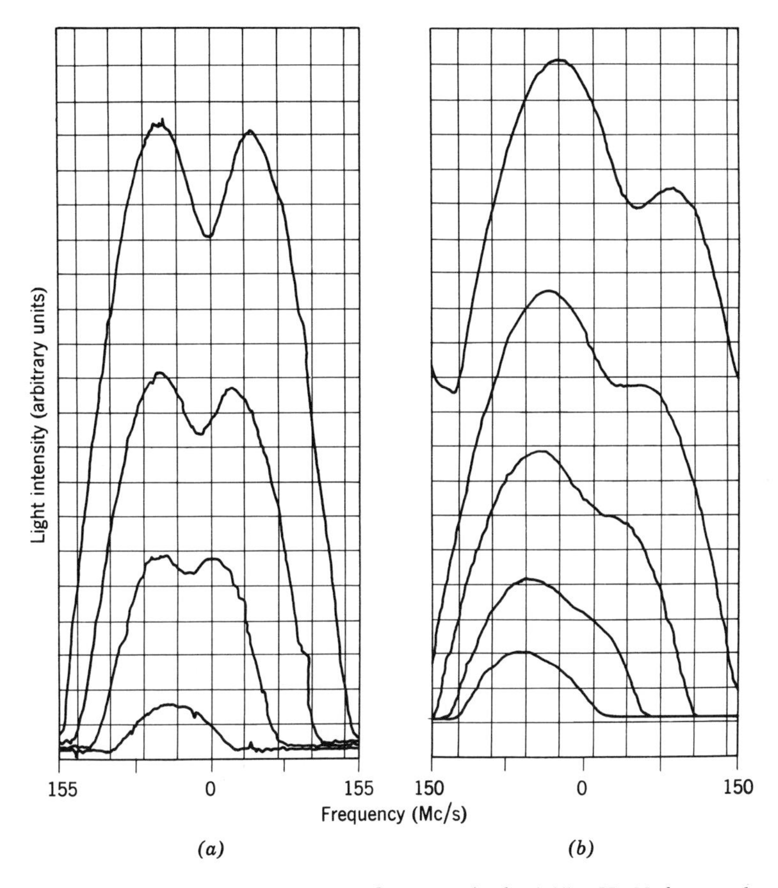

Figure 33. Illustrating power output versus frequency in the 1.15 μ He-Ne laser as the gain is varied. (*a*) The results for pure ^{20}Ne isotope. (*b*) The results for natural neon. (Reproduced by permission from [34].)

if all conditions are optimized for observing a power dip, however, the dip is rarely as deep as one-half the Gaussian due to many complications. One of the principal reasons is the ratio γ_{ab}/γ'_{ab} in (67). This ratio is typically of the order of 0.1 to 0.2 in helium-neon lasers, but it may be slightly larger in some other types of laser. We should also note that the power dip exists because of the presence of a standing wave in the laser resonator, and special resonators that are designed to operate with traveling waves rather than standing waves, such as those that are designed in the form of rings, will not demonstrate a power dip.

The power dip is centered in the middle of the Doppler profile and thus, when it exists, demonstrates the fact that radiation into the laser mode comes from slow moving atoms. After the initial demonstration of the dip, it was quickly realized that it could be used to define the output frequency of a single-frequency laser to a higher degree of accuracy than could be obtained merely by attempting to center on a relatively broad Doppler curve. The curves of Figure 33, for instance, suggest that an improvement of as much as a factor of 10 may be achieved by sensing the minimum of the power dip, as opposed to what could be achieved by sensing the maximum of a Doppler curve. The greatest application for stabilized lasers, however, is at the visible line at 6328 Å, and unfortunately conditions on this line are not as favorable for obtaining a well-defined power dip as are conditions on infrared lines.

Visible helium-neon lasers have been built, and in fact are available commercially, that employ the power dip to stabilize the resonator so that its frequency is at the center of the power dip. However, the power-versus-gain characteristic that they have to work on is not nearly as ideal as that shown in Figure 32, and in fact it is more nearly approximated by the curve of Figure 34. Such a laser has to be short, typically 10-cm resonator length, and the peak unsaturated gain G is of the order of 5%. The losses, including mirror transmission, are unavoidably of the order of 1% and thus the relative (gain/threshold) ratio is about 5, a marginal number. The Doppler width δ is 1020 MHz, and the natural width γ'_{ab} that needs to be inserted into (67), unfortunately limited by the relatively frequent soft collisions rather than by the hard collisions, is about 200 MHz, or only about one-fifth the Doppler width. When all these factors are taken into account, the extent of the power dip is only 7 to 10% of the peak available power, even in the best cases; and if the laser deteriorates, the fractional dip becomes even less. The ability to stabilize a servo loop on information obtained from the maximum or minimum of a characteristic curve depends on the second derivative of the curve. It can be shown quite easily that the second derivative of the power dip shown in Figure 34 is not particularly greater in absolute

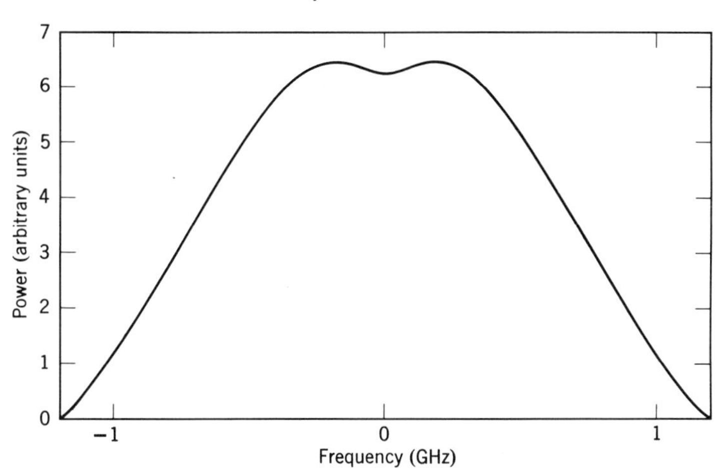

Figure 34. Calculated power output for a small helium-neon laser. Assumed gain/threshold ratio is 4:1.

value than the second derivative that could be obtained from the Doppler curve itself if no dip were present. Thus stabilized single-frequency helium-neon lasers that stabilize on the power dip do so only because the power dip is unavoidably there, not because it provides any real advantage. Nevertheless, the dip is used in this manner because it provides one of the most convenient methods of stabilizing a laser frequency to accuracies of the order of parts in 10^8 over long periods of time. Such lasers have also been made to operate on other visible lines in the helium-neon system (6118 Å, 6401 Å and 7305 Å), but since the gain on these other lines is less than that of 6328 A, the corresponding power dip is even less well-defined than is the case with 6328 Å.

4. Schemes for Stabilizing the Frequency of a Single-Mode Laser

The frequency stability of a single-frequency laser is determined entirely by the stability and length of its resonator. For use as a heterodyne local oscillator source, or for long-distance interferometry, the laser frequency should be stabilized to better than 1 part in 10^6, or, in fact, to 1 part in 10^8 if extensive heterodyne work is to be done with it. If the resonator is 10 cm long, the frequency will change by 5 parts in 10^8 due to thermal expansion alone, if the temperature changes by as little as 0.1° C. For this reason, some sort of active stabilization system appears to be essential in any stable single-frequency laser.

A detailed review of methods that can be used to stabilize a single-frequency laser has been given by White [72]. Briefly, we review the salient points here. A laser can be stabilized from a servo signal that is derived either from the laser itself, making use of information derived from slight changes in cavity length, or it can be derived from an external reference cell which may or may not represent an oscillating laser. Thus all schemes for stabilizing lasers appear to fall in one of the following three classes:

1. The laser is stabilized by providing a small sweep signal, which is phase detected and returned to the control element (for example a piezoelectric mounting on one resonator mirror) to correct the resonator length.
2. Two or more lasers are used, one of which is swept to provide control information and one of which is stabilized by the control signal.
3. The laser beam (or a part of it) is passed through a passive external resonance cell, which is not a laser and which may actually have resonant absorption instead of induced emission; the changes in amplitude of the laser beam caused by the resonance cell are then used to actuate the servo system.

The first class—the self-stabilizing laser—has been available commercially up to the present time. It represents the simplest type, in terms of the overall number of components in the package, and thus is better suited to commercial development than the other lasers, which may require more than one discharge tube or even more than one highly stabilized cavity. A typical scheme for stabilizing a laser in the center of the power dip (Figure 35) applies a rapid ac sweep voltage superimposed on a dc bias to a piezoelectric element, the piezoelectric element constituting part of the mounting of one resonator mirror. The effect of the ac sweep field is to sweep the laser frequency by small amounts. A portion of the laser output is detected by a photocell, and the output of the photocell is passed through a phase detector which compares it in amplitude and phase to the original sweep field. The output of the phase detector is thus approximately equivalent to the derivative of the power curve shown in Figure 34. The phase detector output, suitably filtered to remove all but low-frequency components, is imposed upon the piezoelectric element as part of its dc bias. When the resonator length is correctly set so that the laser frequency is at the center of the power dip, the derivative is zero and there is no output from the phase detector. If the frequency drifts one way or the other, a corresponding derivative of the signal is generated. This acts on the piezoelectric elements to correct the length back to the center of the power dip.

If the power dip did not exist, the same scheme could be used to center on the Gaussian Doppler curve itself, except that the derivatives

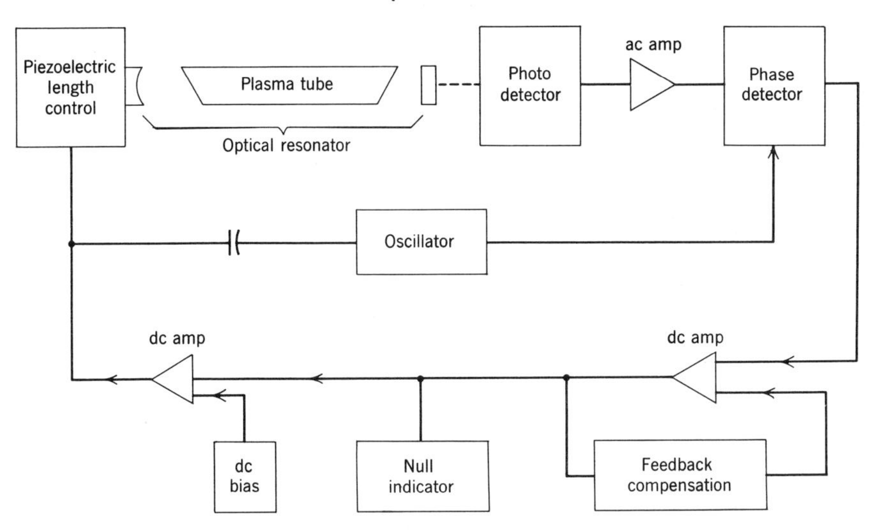

Figure 35. Block diagram of laser stabilization scheme.

would have opposite signs on either side of line center and one would have to interchange the sign of the input to the piezoelectric element. A problem in the use of the power dip as a stabilizing point in the manner just described is the fact that the servo loop is stable only when the laser frequency is within the power dip itself. If the frequency should, for any reason, drift outside the power dip, the sign of the derivative becomes wrong and the servo acts as an unstabilizing element, tending to drift or oscillate rapidly in the region at the edge of the Gaussian (where the laser output "jumps" from one mode to the next mode). For this reason, some sort of test is usually necessary in setting up a single-frequency laser of this type to make sure that the feedback loop is correctly locked in the center of the power dip (Figure 36).

Although other schemes for employing laser output have been proposed, they have generally proved unsuccessful. This is particularly true of schemes that involve sensing of absolute amplitude changes as a function of position along the Gaussian curve, because long time variations in the power output of any laser occur as the optics become more lossy with time.

A further problem in the use of self-stabilizing lasers is that the power dip depth and position are in themselves more susceptible to effects of pressure broadening than are some other characteristics of the output curve. In particular it has been shown that the operating frequency of a self-stabilized laser centered on the power dip may be a function of

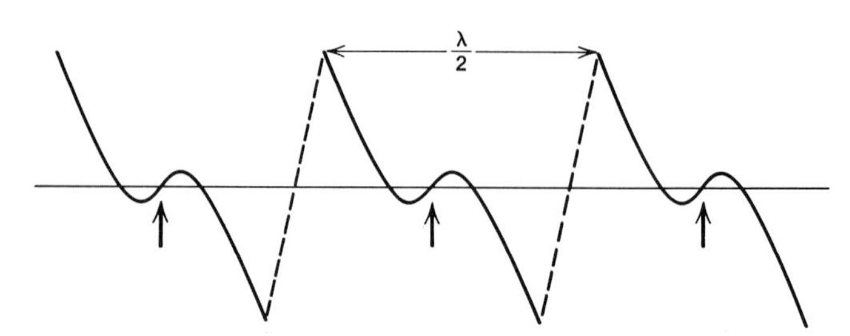

Figure 36. Output of the phase detector (derivative of the power curve) for three half-wavelengths variation in mean length of the resonator. Correct locking of, the stabilization system on the power dip is at any of the points marked by the bold arrows. However, the servo system can also lock at points where the dashed lines cross the axis.

discharge current under some conditions [73, 74]. When this and other factors are taken into account, the long-term stability to be expected from a laser of this type operating at 6328 Å is generally of the order of parts in 10^7, although further improvements may make possible stabilities of the order of parts in 10^8.

Schemes employing two lasers, one to help stabilize the other, have been discussed in the literature, and it has been shown that the potential long-term stability that can be obtained should be very high. Laboratory experiments designed to measure drift and signal-to-noise ratio agree with the magnitudes of the predictions [75, 76], although problems still exist because of the need to stabilize two or more resonant cavities. This, plus the fact that systems of types 1 or 3 are generally simpler in conception, has limited the use of two lasers to stabilize each other to laboratory experiments.

Schemes for stabilizing a helium-neon laser on a passive atomic resonance require an additional discharge tube similar to the laser tube, containing a discharge of neon or helium-neon mixture. If the tube is filled with pure neon, there will be a net absorption of the 6328 Å light, generally of the order of 1% of its initial value, and if filled with a helium-neon mixture there will be a net negative absorption or stimulated emission of the same order of magnitude. Either type of discharge cell can be used, if correct attention is paid to the sign of the output signal.

One of the best methods for making use of a passive cell is that employed by White [72], as shown in Figure 37. In this scheme a magnetic field is employed to produce a Zeeman splitting of the passive resonance, and sensing of the two circularly polarized components in the longitudinal Zeeman effect is accomplished by passing circularly polarized laser light through the cell. In the scheme shown, an electro-optic quarter wave plate is used to switch the laser output alternately between right-handed and left-handed circular polarization. The output of the phase detector is detected at a frequency synchronous to that of switching the polarization. When the laser output frequency is at the center of

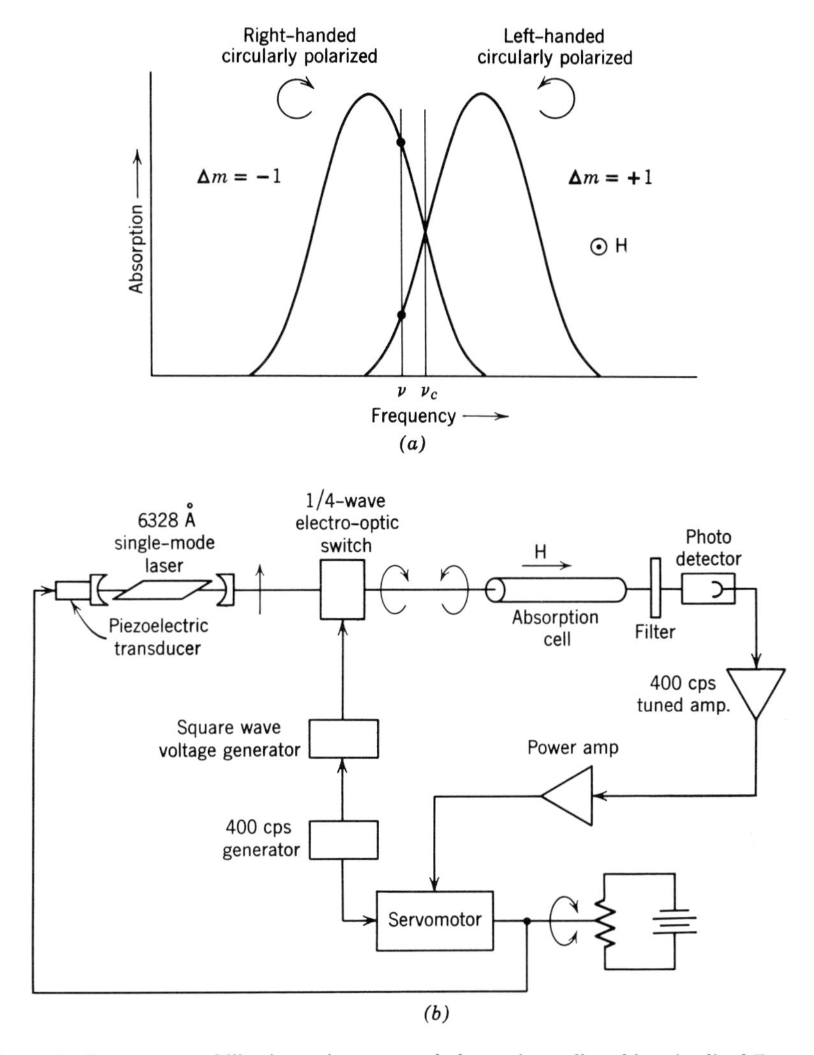

Figure 37. Frequency stabilization using external absorption cell and longitudinal Zeeman effect. (*a*) Relative absorption of right- and left-handed circularly polarized laser light at frequency ν as a function of deviation from correct frequency ν_c. (*b*) Block diagram of stabilization scheme. (Reproduced by permission from [72].)

the (unshifted) absorption line of the passive cell, the absorption from both circularly polarized Zeeman components is the same and the net output of the phase detector is zero. If the frequency shifts for any reason, the signals become unequal and there is a net output from the phase detector. The signal characteristic out of the phase detector is quite similar to that of an FM discriminator in a radio receiver, and the philosophy of the detection process is much the same. The scheme has several important advantages, chiefly that it depends on the line center of a natural Doppler curve that may be controlled independently of the Doppler curve of the laser itself. In addition the signal-to-noise ratio of the system is very high, thus promising a high degree of long-term stability. The short-term stability of such a scheme may not be as good as that of a self-stabilized laser and it may be necessary to introduce an additional feedback loop containing high frequencies, based on the laser output itself, to control rapid variations.

The chief promise of schemes using passive cells is in their long-term absolute stability and possible reference to atomic resonant frequencies. The possibility of using mixtures in the reference cell other than good laser mixtures allows for a high degree of control over pressure- and current-dependent shifts. Measurements directed toward this end have been made by White [77].

Study of schemes for stabilizing lasers is continuing. Recently there has been much work on mode and polarization competition effects in lasers operating in one or two modes and of the possibility that these effects might be used for stabilization [78]. If internal mirrors are used, the output polarization may be a strong function of position of the laser frequency under the Doppler curve, and this also shows promise for use in stabilizing the laser [79]. The reader is referred to the paper by White [72] for additional details.

It should be pointed out, in addition, that the discussion here has been in regard to stabilization of lasers that can operate in only one temporal mode at a time, by virtue of the mode spacing and gain profile. Stabilization of naturally multimode lasers is also possible, and some of the schemes for doing this are discussed in Chapter 4, Section D.

5. Ring Resonators

A special type of resonator that has properties somewhat different from those that have been discussed is the ring resonator. Although lasers employing such resonators are used only in specialized experiments, it seems worthwhile to present a brief review here.

The basic plan of a laser employing a ring resonator is shown in Figure 38. Basically, the optical path consists of a closed path in the form of a polygon; a square is shown but a triangle would involve one less mirror with its attendant losses. A laser plasma tube is put in one or more of the sides to provide gain, and some output means is provided. The temporal modes of this resonator are traveling waves in both the clockwise and counterclockwise directions, and the wavelengths are determined by the requirement that there must be an integral number of wavelengths in a complete circuit of the ring. If L is the perimeter of the ring, then the temporal modes in each traveling wave direction are separated by c/L in frequency, not $c/2L$ as in a standing wave resonator.

The spatial modes of a ring resonator are the same as those of a standing wave resonator, if one of the corner mirrors is considered to stand for both of the end mirrors of the standing wave resonator. In order to maintain stability against diffraction losses, at least one of the corner mirrors must be curved to match spherical wave fronts approaching the mirror. A detailed discussion of the spatial mode control in ring resonators has been given by Rigrod [80].

The most interesting property of ring resonators is that the apparent frequency difference between temporal modes of the clockwise and counterclockwise waves is proportional to the rate of rotation of the ring. The reason for this can be seen from the following argument. Con-

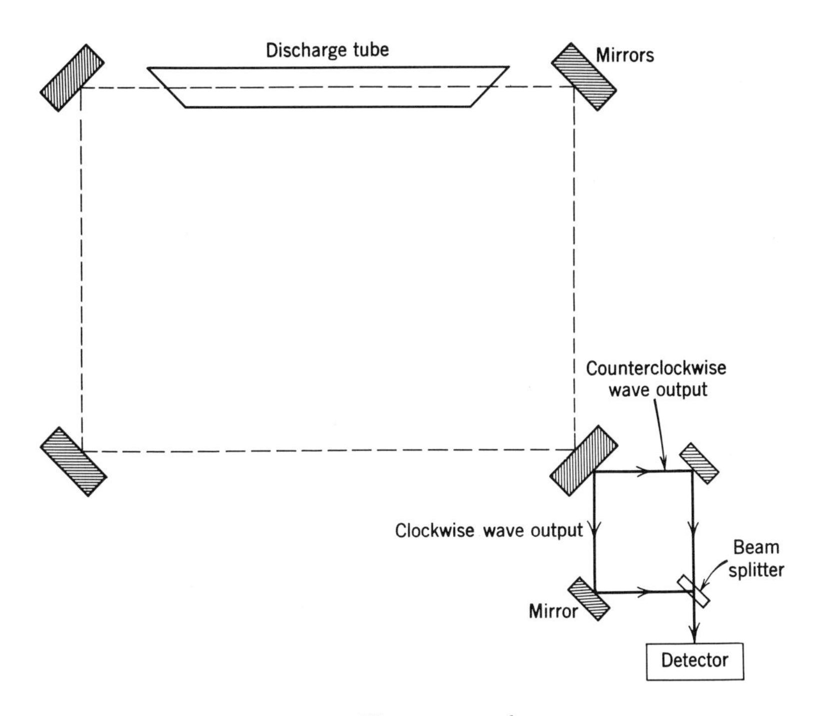

Figure 38. Ring resonator laser.

sider a nonrotating ring containing two oppositely directed modes with the same temporal mode number q—this is equivalent to a standing wave of mode number $2q$. Now, if the resonator is rotated, and ignoring distortions caused by the shape of the polygon, the standing wave remains the same—light travels at speed c in both directions—but the output device samples different parts of the standing wave as it moves. This is equivalent to observing equal and opposite Doppler shifts in the two traveling waves. If the output device combines the output of the traveling waves interferometrically, then a photocell observing the combined output will generate a beat note proportional to the rotation rate of the ring, provided that the photocell is rotating along with the ring.

The exact relationship between observed beat frequency and the rotation rate of the ring depends in detail upon the ring geometry. If the ring is in the form of a square, then it has been shown [81] that the beat frequency $\Delta\nu$ is given by

$$\Delta\nu = \frac{\nu\,\Omega\,D}{c}\cos\,\gamma\,, \tag{68}$$

where ν is the unperturbed (zero rotation rate) frequency of the laser output, Ω is the rotation rate in radians per second of the laser, D is the length of one side of the square, and γ is the angle between the rotation axis and the normal to the square. Experimental verification of (68) has been given by Macek and Davis [82] and Cheo and Heer [83], as well as by subsequent investigators. For a detailed and general discussion of the theory, the reader is referred to the paper by Heer [81].

The applications that are of greatest interest are, of course, those that require observation of very low rotation rates, often with a relatively small ring, so that the desired beat frequency is very low. Unfortunately, as the beat frequency approaches zero, there is a tendency for the traveling waves to lock together into a standing wave that is fixed relative to the resonator structure, giving effectively zero beat frequency. The locking effect is caused by energy that is coupled from one traveling wave to the other by a structure that is fixed relative to the ring; media that can provide such coupling include diffuse reflections and light-scattering from mirrors, Brewster angle windows, and detector surfaces. In experiments where the greatest care has been taken to avoid coupling effects between traveling waves, locking still occurs when the expected beat frequency is of the order of 10 Hz. The locking effect can be eliminated by providing an initial offset between traveling wave frequencies; for example, by operating with traveling waves having different temporal mode numbers. Various means for providing the initial offset on a practical basis have been discussed in the literature.

CHAPTER 4

THE GAS LASER CONSIDERED AS A LIGHT SOURCE OR LOCAL OSCILLATOR

In this chapter we take a more abstract view of the gas laser than we have taken before. In earlier chapters the concern has been primarily with details of operating lasers as they actually exist; here we consider the properties of a gas laser beam, assuming that all problems regarding the construction of the laser have been solved and that the laser can now be considered to be a classical oscillator generating a beam of light waves. The first three subsections are concerned purely with the properties of the light beam itself. The last one has to do with methods of control of the light output from the laser by means other than control of the gain, the topic of the previous chapters.

A question may arise as to how closely actual physical gas lasers approach the "ideal" lasers that are assumed in the following sections. Experiments with highly stable 1.15-μ lasers suggest that a helium-neon laser can be made to have a noise level quite closely approaching the ideal, under suitable conditions of low pressure, low current density, and RF excitation. Most other lasers are somewhat noisier than the ideal, but conditions can apparently be found for any given laser in which the noise level can be reduced to a point where it is of no practical importance. The ideal Gaussian spatial mode considered here is also achievable, but occasionally at the expense of output power because it requires severe aperturing in order to suppress higher-order modes. Small amounts of doughnut mode can often be tolerated in a laser beam that is intended to be operated only in lowest-order spatial mode, without particularly deleterious effects, because the focused spot size of the doughnut is only slightly larger than that of the lowest-order mode. Many so-called single spatial mode lasers probably do in fact contain substantial amounts of energy in the doughnut mode. This factor is not important optically; it is important only when heterodyne experiments are to be performed with the laser beam and it is found that the higher-order mode generates beat frequencies that are different from those generated by the lowest-order mode.

A. PROPERTIES OF AN "IDEAL" GAUSSIAN LASER BEAM

These properties, as they apply to this section, are summarized in Table 7. All of these points have already been discussed, except possibly number 3, the question of the type of noise recorded by a good photodetector looking directly at the laser beam output. The implication of "white" noise is that the time variations in energy distribution of the beam are only those that are required by quantum statistical theory. Any noise caused by macroscopic plasma fluctuations or other effects to be discussed will probably generate noise having a peculiar frequency distribution, discrete frequencies, or other large deviations from a uniform noise distribution.

TABLE 7
"Ideal" Laser Beam

1. Plane or spherical wave.
2. Gaussian intensity distribution.
3. Detector sees "white" noise.
4. Single frequency
   $\Delta\omega \approx 1$ Hz

1. Propagation as a Plane or Spherical Wave.

The equation of Boyd and Kogelnik (56) contains, in principle, all that one needs to know about the propagation properties of a laser beam in the lowest-order spatial mode. However, the equation as originally expressed applies primarily to spatial variations within the laser cavity, and it is desirable to expand the equation somewhat for other purposes.

An ideal laser beam must have either a plane or spherical wave front. If it has a plane wave front, it is at its point of minimum diameter (i.e., its "focus") and must change to a diverging spherical wave after propagation through any distance. If it is a converging spherical wave, it will become a plane wave at the one point in space at which its diameter is

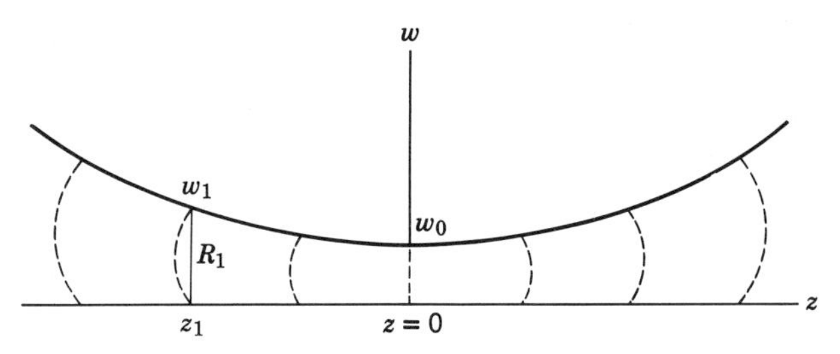

Figure 39. Gaussian beam propagation showing w as a function of z. The variation of R is sketched in schematically.

a minimum. Geometrically, the properties of the wave front must be symmetrical about the "focus" unless the beam is passed through optical elements that change the radius of curvature of the wave front.

These properties are illustrated in Figure 39. Given a known wavelength λ; if the radius of curvature of the wave front, R_1, and the spot size, w_1, are known at one point in the propagation path, then the wavefront characteristics are determined for all points on the propagation path. First, the given point is located at a distance z_1 from the "focal point" or point of minimum beam diameter. This distance is given by

$$z_1 = \frac{R_1}{1 + \left(\dfrac{\lambda R_1}{\pi w_1^2} \right)^2}. \tag{69}$$

The spot size at the focal point, w_0, is given by

$$w_0^2 = \frac{w_1^2}{1 + \left(\dfrac{\pi w_1^2}{\lambda R_1} \right)^2}. \tag{70}$$

Then, w and R can be calculated at any other distance z from the focal point. This is given by

$$w^2 = w_0^2 \left[1 + \left(\frac{\lambda z}{\pi w_0^2} \right)^2 \right], \tag{71}$$

$$R = z \left[1 + \left(\frac{\pi w_0^2}{\lambda z} \right)^2 \right]. \tag{72}$$

It is sometimes convenient to know the "equivalent confocal radius" b, the length of a confocal resonator whose central value of w equals w_0. This is given by

$$b = \frac{2\pi w_0^2}{\lambda}. \tag{73}$$

At $z = b/2$, $w = w_0 \sqrt{2}$.

These simple propagation properties of the Gaussian beam have been displayed in a particularly convenient fashion in a chart by Collins [84], which is now commonly known as the "Collins chart." It is analogous in form to the Smith chart familiar in microwave technology, and commercially available Smith charts may be used as Collins charts with their coordinates suitably labeled. Figure 40 is the chart originally given by Collins. It will be observed that the chart consists of circles that are tangent to the origin and circles that intersect the origin (in Collins' notation his quantity x_0 is equivalent to our quantity w). The circles that intersect the origin correspond to constant values of z, whereas the circles that are tangent to the origin correspond to constant values of the equivalent confocal resonator b. Since propagation through space is always defined by a particular value of b, such propagation is indicated on the chart by projection as a portion of a given curve of constant b. The x-axis of the chart corresponds to the value $z = 0$ and values plus or minus can be read either up or down from that axis of the chart. Examples of how this may be used are shown in Figure 41.

A particularly convenient feature of this representation is the fact that the wave front radius of curvature R is indicated by its inverse on the y-axis; thus y is equivalent to $1/R$. This makes the chart particularly convenient to use with thin lenses and other simple focusing elements. Consider an incoming wave front having a radius of curvature u, which is converted by the lens to an outgoing wave front having a radius of curvature v. The well-known lens formula

$$\frac{1}{u} + \frac{1}{v} = \frac{1}{f} \tag{74}$$

states simply that the sum of these inverses is a constant. In the Collins chart these inverses are indicated directly on the chart and the constant $1/R$ corresponds to a fixed length in the y direction. Thus a simple lens has a very simple representation on the Collins chart. In traversing through free space one merely plots propagation along a curve of con-

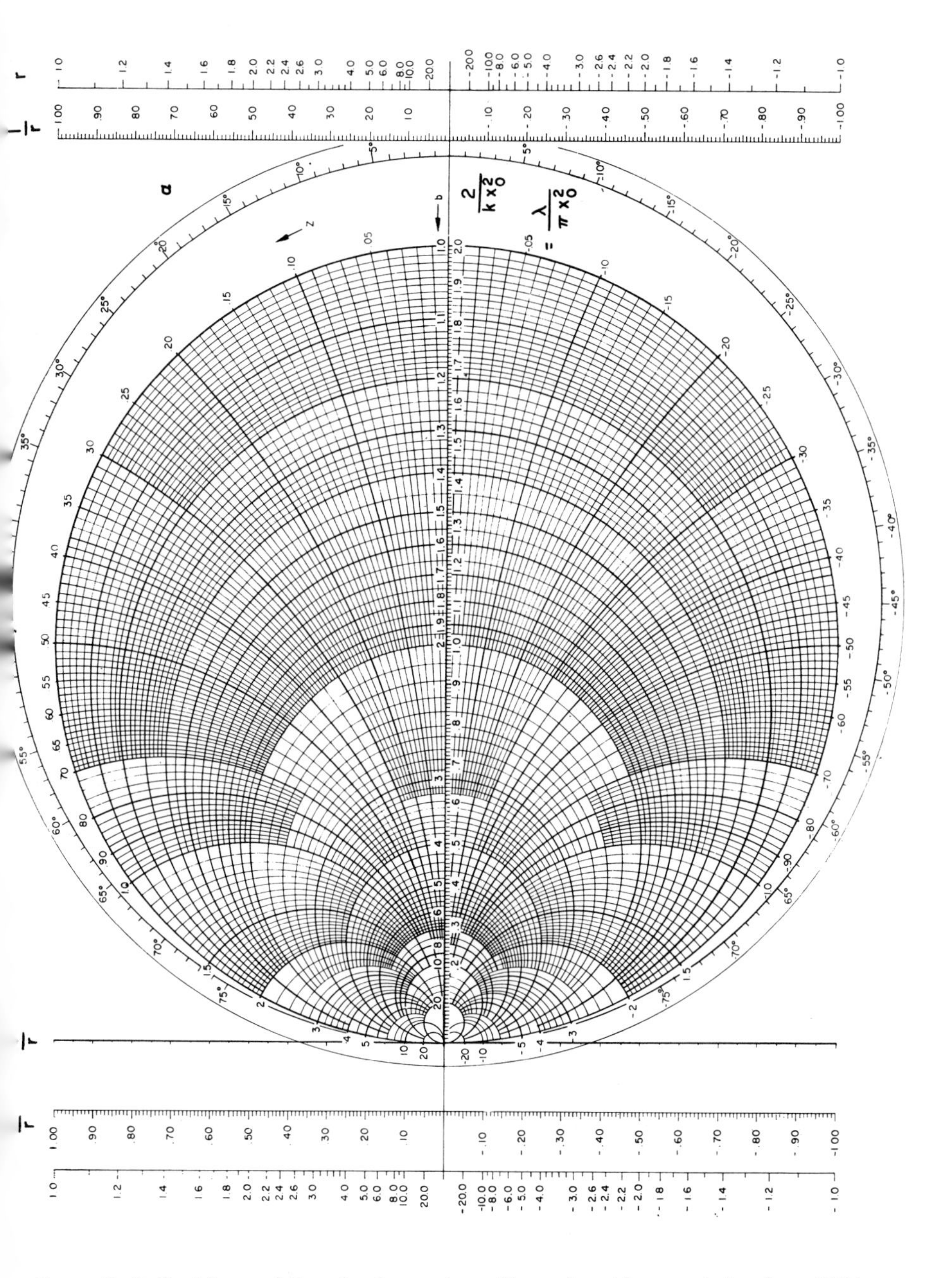

Figure 40. Collins' form of Gaussian beam chart. (Reproduced by permission from [84].)

stant b. Then, at the point at which the lens is placed, the inverse focal length of the lens $1/f$ is included as being equivalent to propagation along a straight vertical line of length $1/f$ on the chart, following which propagation in free space proceeds along a new b circle, as before. This is illustrated in Figure 41.

The Collins chart thus provides a convenient method of getting "from here to there" when Gaussian beams are passed through an optical system containing thin lenses. Figure 41 shows such a procedure in which it is assumed that the wave front initially corresponds to the point A, and it is desired to refocus after a certain propagation distance to point D on the chart. The figure shows how this can be accomplished with one thin lens whose focal length is equivalent to the straight line BC. Obviously, this simple problem could be done without the chart, and there are alternate methods of performing the same solution (for example, the lens B' C'), but the technique may be particularly convenient where more complicated systems are involved. For example, given two endpoints on the chart and several lenses of known focal length, one might consider constructing straight vertical lines of appropriate length corresponding to the lenses and manipulating them by hand upon the chart until a closed curve is formed consisting only of portions of b circles and the vertical straight lines. When this solution is found, the corresponding propagation distances between lenses and

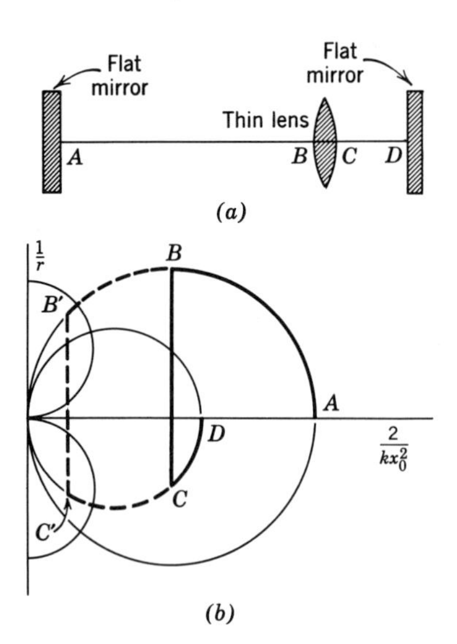

Figure 41. Use of a Gaussian beam chart with thin lenses. (*a*) Resonator. (*b*) Graphical representation of a resonator.

their locations can be determined from the chart, as well as the spot sizes w at all points in the propagation path. Alternatively, one may use the chart to manipulate a propagation path such that the spot size w does not exceed a certain size, and so on.

2. Effective Brightness of Source and Focused Spot

All Gaussian beams are in a sense equivalent since they do not change their shape through free space propagation. This property has this important corollary: if a Gaussian beam can be formed by a source of a given radius, for example w', then any other Gaussian beam of the same wavelength, whatever its present diameter or curvature, can be focused by means of suitable optics down to a point of the same radius w'. This interconvertibility is perhaps one of the most useful properties of laser beams and is definitely a property of coherent light that is not shared with light from incoherent sources. In particular thermodynamical arguments require that incoherent light coming from a source with a certain brightness can never be focused to an image that has a brightness greater than that of the original source. In the case of a laser, however, source brightness cannot be defined thermodynamically, but instead one defines the beam properties in terms of the Gaussian beam parameters (Figure 39) and the total energy in the beam; the effective brightness of this source then depends upon the smallest image that one can form with a Gaussian beam of this wavelength. Simple arguments based on electromagnetic theory suggest that such a focused image can never be smaller than approximately one wavelength in diameter. We now go into this in more detail.

It is well known in classical optics that, given an optical system defined by a particular F-value or F-number, the definition of the image is approximately inversely proportional to the F-number. In the case of a Gaussian beam, when the F-value of the optical system is greater than about F-4, this argument is certainly true and can be used to define both the effective focused spot size and the central brightness of the spot, where the spot size is calculated from (69) and (70). It is necessary to redefine what is meant by F-number of a lens for this case, since it is usually defined in terms of the angle subtended by the aperture. For present purposes, we define the F-number as shown in Figure 42, that is, in terms of the angle subtended by the $1/e$ points of the Gaussian intensity distribution. In order to transmit the Gaussian faithfully and not distort it at the focal point, the actual F-number of the lens in terms of included angle between points on the aperture must be somewhat greater than the Gaussian equivalent value of F-number (in fact, it should be at least three time greater), and this means that spot sizes calculated by this value of F-number will be considerably smaller than the corresponding diameter of the Airy disk that is obtained if a uniform plane wave is used to illuminate the same aperture. This fact should be borne in mind in comparing the two cases, although for present purposes the definition of F-number given in Figure 42 is used because it provides particular simplicity to the equations.

If we assume that an incident laser beam has a total energy I_0, we can calculate the equivalent brightness of a focused spot as follows. Assume that the lens has a focal length equal to 1 (this assumption simplifies the derivation but has no other consequence). We now define the incident Gaussian beam in a slightly different manner from that employing the spot size w, which came about from a derivation in which w referred to field strength. Here we prefer to refer to intensities, and we define the mean width of the Gaussian as δ, where

$$\delta = \frac{w}{\sqrt{2}}. \tag{75}$$

Since the radius of curvature of the wave front converging from the lens is one, the equivalent F-number f is given simply by

$$f = 2\delta. \tag{76}$$

By substituting (75) and (76) into the propagation equations (69) and (70) with $R = 1$ we can determine the corresponding value of w_0 and therefore of δ in the focused spot in the focal plane. The actual intensity in the focused spot at the origin is then obtained merely by integrating over the total energy of the focused spot and setting it equal to the incident energy. The result of such a derivation follows. If $I_{central}$ is the energy per unit area at the center of the focused Gaussian, then we have

$$I_{central} = \frac{\pi^2 I_0}{\lambda^2 f^2}, \tag{77}$$

where $I_{central}$ is *energy per unit area* but I_0 is *total* incoming energy. A numerical example may be convenient here. Suppose that the total incident power is 1 mW and that the equivalent F-number (defined as in the preceding paragraphs) is F-10. By substituting the wavelength for 6328 Å light into (77), the central intensity of the focused spot is 0.25 mW/μ^2, or, in more common units, 250 W/mm^2.

For further investigation into the problem, it is desirable to introduce some additional units of measurement. When working with reasonably high F-number systems, it is well known that the spot size is inversely proportional to the F-number, as mentioned, and also that the depth of focus of an optical system is proportional to the square of its F-number.

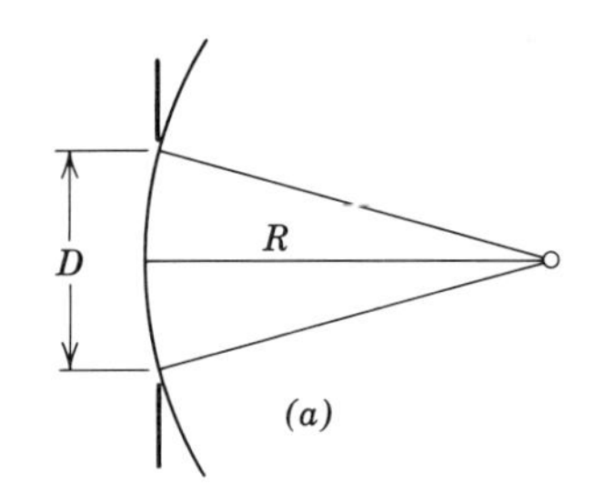

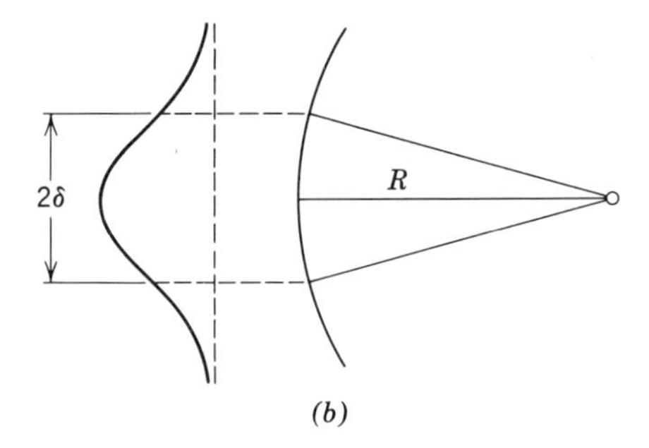

Figure 42. Definition of F-number. (a) Classical definition for uniform wave, $F = R/D$. (b) Definition for use with Gaussian beam, $F = R/2\delta$.

This suggests the possibility that a new system of linear coordinates may be used to describe behavior of the light wave in the vicinity of the focal plane, in which the results are independent of F-number but can be applied quickly to any practical system where the F-number is known. Such a system is used extensively by Born and Wolf [25], and we apply it with some modifications here.

Since the focused laser beam has symmetry about the optic axis, we need only two coordinates, one along the axis and one perpendicular to it. The coordinate along the axis we define as u and the one perpendicular to it as v. If the corresponding linear distance coordinates are z and r (z being along the optic axis) the relationship between u, v, z, and r is given as follows:

$$u = \frac{\pi z}{2\lambda F^2}, \tag{78}$$

$$v = \frac{\pi r}{\lambda F}. \tag{79}$$

The focal point is at $u = 0$, $v = 0$. In terms of these coordinates, the propagation properties of the Gaussian beam in the vicinity of the focal plane have a very simple form,

$$I(u,v) = \frac{I_0}{\pi(1+u^2)} \exp\left[-v^2/(1+u^2)\right]. \tag{80}$$

It is also of interest to specialize this formula to the specific case of the focal plane ($u = 0$) and the optic axis $v = 0$. For these we have

$$I(u) = \frac{I_0}{\pi(1+u^2)}, \tag{81}$$

$$I(v) = \frac{I_0}{\pi} e^{-v^2}. \tag{82}$$

Since one can substitute z or r for u or v very quickly in these formulas, they provide a convenient method of determining the actual physical spot sizes at any point in the propagation path.

A direct comparison of the spot size and intensity for a focused Gaussian and for a focused uniform wave is also of interest. A uniform wave focuses to an Airy disk, (8). In terms of dimensionless coordinates, u and v defined by (78) and (79), the first dark ring of the Airy disk occurs at $v = 1.22\pi$. For a given value of F, a comparison of the focused Gaussian and the focused Airy disk is shown in Figure 43. Again, we remind the reader that the definition of F-number in the case of a focused Gaussian must be taken with caution inasmuch as it requires that the lens be large enough to include significant portions of the wings of the Gaussian.

If the focusing system has a very small F-number—approximately less than F-2—then the results just given have to be modified somewhat due to the vector nature of the electromagnetic theory. Although propagation of a plane-parallel wave in the z direction implies that no z component of polarization (of either the electric or magnetic fields) can be present in the traveling wave, this does not necessarily mean that such components cannot be present in the vicinity of the focal plane of a sharply focused beam. An F-1 lens subtends an angle of approximately 60°, or about 30° from the optic axis. Consider an incoming spherical wave of amplitude E subtending such an angle. The propagation of the wave at any point on the sphere is toward the focal point, and the polarization of the wave at any point is always normal to its propagation. At 30° from the optic axis, however, this implies that the propagated wave does have a component of polarization E_z in the direction of the optic axis to the following extent:

$$|E_z| = |E| \sin\theta. \tag{83}$$

Thus as the wave converges to the focal point it is capable of generating, by interference effects, components of polarization that are parallel to the optic axis as well as those that are perpendicular.

A study of the focusing properties of electromagnetic waves in optical systems of low F-number, making use of rigorous solutions of Maxwell's equations, was begun by Wolf and his co-workers [85]. In particular a paper by Boivin and Wolf [86] shows in great detail the intensity distributions of the vector components of the electromagnetic field in the vicinity of the focal point, when a plane wave of uniform intensity is focused through an optical system having an F-number of approximately F-1. These results, plotted in terms of contour maps showing contours of equal amplitude (or intensity), can be compared directly with similar charts presented in Born and Wolf [25], which show how a scalar wave focused by a system of higher F-number converges to form the Airy disk at the focal plane.

The results of Boivin and Wolf show that the focusing of a plane wave by an F-1 system also gives rise to rings in the focal plane, but there are other more complicated effects caused by the vector nature of the field. In particular, the z component of polarization (component parallel to the optic axis) shows up in the form of two lobes located in the focal plane and symmetrically about the focal point. For the E vector, these

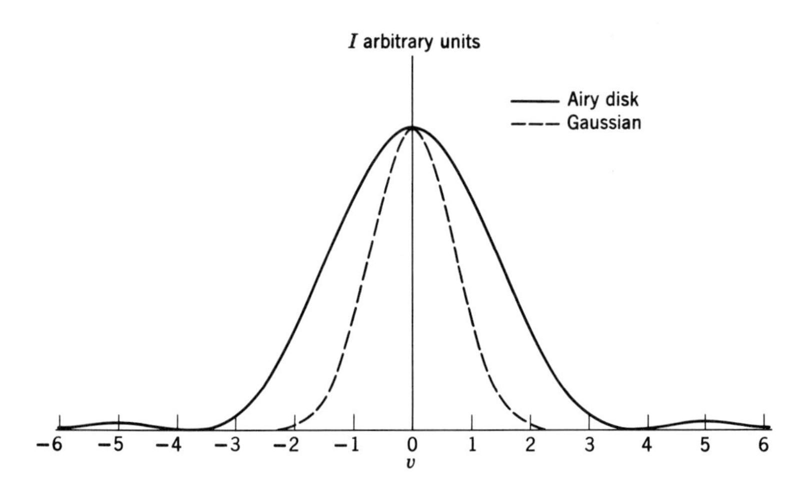

Figure 43. Comparison of focused Gaussian and Airy disk according to definitions of F-number.

lobes lie along the axis defined by the electric field polarization; for the H vector, corresponding lobes of H_z field lie along the corresponding axis at right angles to the axis containing the E_z lobes. These patterns are strikingly similar to those generated by the induction and transition fields of an electric dipole antenna. This result is not surprising since radiation from such an antenna gives rise to diverging spherical electromagnetic waves, and reciprocity theorems suggest that a corresponding spherical electromagnetic wave focused to a point must duplicate some of the field patterns in the vicinity of the antenna capable of generating such a wave.

The results of a similar calculation [87], in which the incident plane wave has a Gaussian intensity distribution, are shown in Figure 44. For this purpose it is assumed that the mean radius of the Gaussian is δ, and that the lens focuses the plane wave directly into a spherical wave at the point at which the sphere intersects the plane wave. (This conversion from a plane wave to a spherical wave is called "aplanatic" by Born and Wolf; other types of conversion are possible, for example, that caused by focusing by a parabolic mirror.) If θ is the angular coordinate in the spherical wave, measured from the optic axis, then the condition F-1 implies that the value of θ that corresponds to δ in the plane wave is approximately 30°. Since this is only the $1/e$ point of the Gaussian, the optical system that generates this converging wave must actually have a considerably greater aperture than F-1 in order to accommodate the wings of the Gaussian.

The calculation that gave rise to Figure 44 assumed that the optical system was able to converge a wave up to $\theta = 90°$, which is not entirely realistic in a physical sense. Nevertheless, the results indicate to good accuracy the shape, intensity, and width of the spot obtained by focusing a Gaussian beam with a low-F-number optical system. First, it will be observed that the peak intensity is decreased by approximately 80% from the intensity that would be predicted if (77) were applied to this case. The mean radius of the spot agrees moderately well with (82), but the spot itself is not circular. The presence of lobes of z-component field, to the extent of approximately 10% of the peak intensity, give rise to the elliptical shape of the spot such that the total intensity distribution is different depending on whether it is measured in a cross section parallel to the axis of incident E-vector polarization or perpendicular to it. It can be seen that the mean width of this focused spot roughly occurs at $v = 1$ and therefore, from (79), at a physical radius of $r = \lambda/\pi$. This is as small a spot as one can expect to obtain, taking into account the previously mentioned fact that an incoming spherical wave must reproduce to some extent the field patterns of the dipole antenna capable of generating the corresponding output wave.

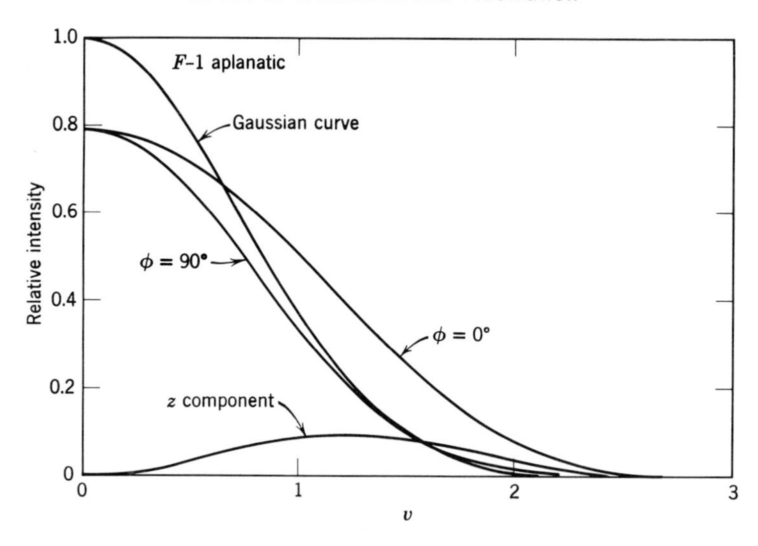

Figure 44. Intensity distribution obtained by focusing a Gaussian beam at F-1. ϕ is the angle between the V-axis and the E-rector polarization of the incident wave. The Gaussian curve itself is shown for comparison.

To summarize, the effective brightness of a laser source giving rise to a diverging wave front, or, correspondingly, the brightness of a focused spot from a wave converging at the corresponding angle, is approximately proportional to the angle of convergence. Equation 77 predicts the central intensity of this focused spot, and the equation can be considered valid up to quite small F-numbers, approximately F-2. Below F-2 the equation can be considered valid to first order, with corrections to be made because the focused spot becomes no longer circular but somewhat elliptical and the central intensity is therefore decreased somewhat from that predicted by (77). The mean radius of the spot in all cases is given fairly accurately by (82).

B. EFFECTS OF TRUNCATION AND ABERRATION ON A FOCUSED GAUSSIAN BEAM

In almost all practical applications involving a Gaussian beam, some truncation of the beam may have to be employed, either because of limitations of physical size or to "clean" the beam of undesired scattered light that shows up at low intensity around the edges of the beam. As a Gaussian beam is truncated more and more, the aperture produces increasingly greater discontinuity at the edges of the beam. When this is focused or otherwise converted into the far-field diffraction pattern, it should be expected to give rise to rings similar to those in the Airy disk. The extent to which such rings might prove troublesome in a given application will, of course, depend on the application, and transmission of a maximum amount of the incident laser energy may also be a consideration in regard to aperturing in some cases.

Detailed calculations have been performed by D. J. Innes and the author upon the effects of aperturing a Gaussian [87]. It appears that an aperture whose radius is two to three times the mean radius of the Gaussian will produce rings that are barely detectable and have an amplitude of perhaps 1% of the peak intensity of the focused spot. An aperture whose radius is three mean Gaussian radii transmits 95% of the incident energy; an aperture whose radius is two mean widths transmits approximately 86% of the total energy. If the aperture is smaller than two mean widths, then the rings become prominent and the intensity distribution in the focal plane becomes similar to that of the Airy disk. When the aperture is equal to the mean Gaussian radius, the focused spot is qualitatively indistinguishable from an Airy disk, despite the fact that the intensity of the incident wave is down considerably at the edges of the aperture from its value at the center. It appears that, to make effective use of most of the optical properties of a Gaussian beam, the beam should not be apertured to less than three mean radii, although aperturing to two mean radii can be tolerated if requirements on the intensity distribution in the focused spot are less stringent. At an aperture of one mean radius or less, the Gaussian beam theory should be discarded and the apertured wave treated essentially as a wave of uniform intensity.

Effects of lens aberrations on transmission of a Gaussian beam have been studied extensively [87] only in the case of spherical aberration, since this is the type of aberration that is most likely to be encountered in an optical system designed expressly for use with laser beams. Spherical and chromatic aberrations are the only aberrations of importance in optical systems designed to be used essentially on-axis, as is the case with most laser systems. The studies that have been made on spherical aberration of a Gaussian apertured to two mean widths indicate that the primary effect of the aberration is to shift the effective focal point to a value between the paraxial and marginal foci, somewhat closer to the paraxial focus than to the marginal focus. The aberration also tends to introduce rings into the distribution in the plane of best focus, but these rings are at very low intensity. In any optical system, spherical aberration increases rapidly as the limiting aperture of the system is increased. Since a Gaussian beam tends to concentrate its energy more along the axis of the beam than is the case with a uniformly illuminated wave, the effect of spherical aberration on a Gaussian beam is, in general, less than it would be for the same optical system used with a uniformly illuminated wave. This is often an important advantage in making use of Gaussian laser beams.

C. NOISE IN THE LASER BEAM AND IN DETECTORS

1. Theoretical Noise Limit

The theoretical noise level of any laser is determined ultimately by the fact that the laser consists of an ensemble of discrete particles interacting with the radiation field, and this interaction is governed by the laws of quantum electrodynamics. In addition the atoms in the upper laser state are capable of losing their energy via spontaneous emission, which is by nature incoherent and thus subject to the usual statistical fluctuations of a black body radiator. This last condition is an important one and sets the laser (or optical maser) apart from radio frequency masers and other coherent RF generators in which the contribution from spontaneous emission is negligible.

The noise fluctuations of an operating laser, assuming that they are only those required by quantum statistics, can be calculated in several ways. A rigorous calculation making use of quantum electrodynamics will not be attempted here. In general, the noise can be calculated first on the assumption that the laser is a single pass untuned amplifier (i.e., merely a column of amplifying medium without end mirrors), then the end mirrors are put on and the oscillation conditions as well as the band pass properties of the resonator are taken into account.

The amplifier noise of an untuned laser amplifier can be considered essentially the "front end" noise of the laser amplified by the remainder of the laser. In the limiting case of absence of other noise sources, this front end noise can be considered merely the contribution of spontaneous emission to the output. The total output, spontaneous and induced, of an excited atom per unit of electromagnetic phase space is given by the well-known quantum mechanical transition probability formula, (12), which we repeat here for convenience:

$$p\,d\Omega = \frac{8\pi^3 e^2 \nu^3 d\Omega}{hc^3} \,|\,\mu\,|^2 (n_\nu + 1)\cos^2\theta. \tag{12}$$

The quantity n_ν is a dimensionless number that might be described as "photons per second per spatial mode per unit frequency range." In addition this must be defined for each of two orthogonal polarizations. Thus the factor n_ν in (12) represents the stimulated emission, caused by photons already present in the unit of phase space, whereas the 1 represents the constant, spontaneous emission.

If the population inversion were complete, a calculation of the noise output, per unit frequency range and per unit spatial mode subtended by the amplifier, could therefore be calculated from (12) without knowing expressly the values of the physical constants involved in the equation. If the value of the amplifier gain G were equal to 2, then the noise contribution is given simply by the spontaneous emission part of (12), that is, by setting $n_\nu = 0$. The noise power for arbitrary gain is given by $(G - 1)$ times the spontaneous emission part of (12). In an actual laser with less than 100% inversion, the spontaneous emission is determined by the upper state population N′, whereas the gain is determined by the population *difference* $N' - N$. All signal-to-noise ratios calculated in this manner or from Table 8 must therefore be degraded by the ratio $(N' - N)/N'$, which, even in high-gain lasers, is not likely to be more than 0.1. The actual signal-to-noise ratio of such an amplifier output detected by a photodetector will be somewhat less than this number and will depend upon the integration time of the detector—the final bandwidth of the complete detection system—since the detector generates discrete photoelectrons and statistical averaging procedures must be used to average over the random distribution of departure times of these photoelectrons.

TABLE 8

Passbands and Noise Levels at 10-dB Gain of Various Gas Laser Transitions. Neutral Atoms Assumed at 400° K.

λ (μ)	Species	Assumed Bandwidth (MHz)	Noise [illegible]
0.6328	Ne	2000	[illegible].9 $\times 10^{-9}$
1.1523	Ne	1100	1.7 $\times 10^{-9}$
3.3913	Ne	370	5.2 $\times 10^{-10}$
2.0261	Xe	250	2.2 $\times 10^{-10}$
3.5070	Xe	144	9.4 $\times 10^{-11}$
0.4880	Ar^+	4200	1.7 $\times 10^{-8}$
0.5145	Ar^+	4000	1.6 $\times 10^{-8}$
0.5682	Kr^+	3000	1.3 $\times 10^{-8}$
10.6	CO_2	72	1.3 $\times 10^{-11}$

For further discussions the reader is referred to the literature. It is worth pointing out in any case that laser amplifiers are intrinsically noisy because the "effective temperature" of the amplifier is determined by spontaneous emission and given by

$$T_{eff} = \frac{h\nu}{k}, \tag{84}$$

(k = Boltzmann's constant)

rather than by the ambient temperature, which is the predominant factor in radio frequency work. Thus the effective temperature of a laser amplifier in the visible may be of the order of several thousand degrees, whereas the temperature of a photoelectric detector which is ultimately used to convert the light signal into electrical signals may be at room temperature or lower. For this reason it is better to detect a weak optical signal directly and amplify it electronically, rather than to attempt to amplify the optical signal in a laser amplifier first. Laser amplifiers may have their uses, but amplification of weak signals is not one of them.

If the laser amplifier is converted into an oscillator by placing a resonator about it, then the amplified spontaneous emission within the laser gives rise to fundamental noise, commonly referred to as "excess noise." A formula given by Gordon [88] for this excess noise can be written as follows:

$$P(\nu)\,d\nu = h\nu\,d\nu\left(\frac{N'}{N'-N}\right)\times\frac{(G^2-1)(1-R)}{(1-R^{\frac{1}{2}}G)^2+4R^{\frac{1}{2}}G\sin^2\phi}, \tag{85}$$

with

$$\phi = 2\pi(\nu-\nu_0)\frac{L}{c}. \tag{86}$$

[illegible] the laser oscillation frequency, L the resonator length, and [illegible]ance at the output end assuming that the mirror at the opp[illegible] reflectance. This formula can be interpreted as being the las[illegible]aneous emission noise amplified and enhanced by the resonator, then degraded by the transmission of the mirror. A study of the formula indicates that, because of this interpretation, the relative noise fluctuation level should be greatest when its bandwidth per mode is narrowest, and this will occur when the laser is operating just barely above threshold. Considerable study has been made of the excess noise in lasers operating either just barely below or just barely above threshold, largely with the intent of determining whether the characteristics of the noise output actually do agree with the behavior predicted from quantum statistics. There is every indication so far that this is indeed the case, if all possible extenuating circumstances that can give rise to modifications of simple statistical distributions are taken into account [89, 90].

Table 8 gives typical numbers for the noise power output of passive amplifiers corresponding to various gas lasers in common use at the present time. All numbers are based on a single pass gain of 10 dB and 100% inversion and must be scaled accordingly to the actual gains and inversions of such lasers. If the laser is used as a passive amplifier, the noise power given in Table 8 is spread more or less uniformly throughout the bandwidth given in the table, which corresponds to the mean Doppler width. In an operating laser having the same gain, the noise output power is approximately the same, although it must be considered to be distributed not uniformly but concentrated in narrow bands about each oscillating temporal mode. Considering the total output power of these lasers, it will be seen that the theoretical noise power is very small and generally unobservable. It is observable only in experiments designed to look for it specifically.

2. Extraneous Noise Sources—Plasma and Mode Interaction Noise

When amplitude modulation noise is actually observed in a laser output, it is generally due not to the theoretical quantum noise but rather to macroscopic effects representing momentary variations in gain or power output of the laser. The two important sources of such noise are variations in plasma density and current and effects of mode interactions in lasers operating under multimode conditions. An additional source of noise, one which is often associated with mode interaction noise, is that caused by microphonic variation of the laser resonator parameters; however, this is essentially a mechanical problem and one not of direct concern here.

Plasma noise can arise because of variations in plasma density, in both space and time, which can exist in the laser discharge. Plasma noise and oscillations as they effect lasers have been studied by numerous investigators, particularly in helium-neon lasers. An especially good theoretical treatment of the factors that give rise to such plasma instabilities has been given by Ingard [91]. This treatment is based on the fact that the negative resistance characteristic of a glow discharge, combined with the impedance variation in the discharge as a function of pressure, makes it possible for the discharge to amplify acoustic waves. Thus, in addition to such acoustic waves as might be provided externally (by coupling to the outside world), there are always noise sources that exist purely because of statistical variations in the density of the gas. The amplitude of the plasma variations, as well as the frequency distribution, is a widely varying function of all the parameters that can enter into the discharge, including current, current density, pressure, discharge tube cross section and length, and composition of the gas giving rise to the discharge.

Figure 45 reproduces curves given by Prescott and Van der Ziel [92] showing variations in the noise in the laser light output as a function of frequency for a particular helium-neon laser. Experiments performed

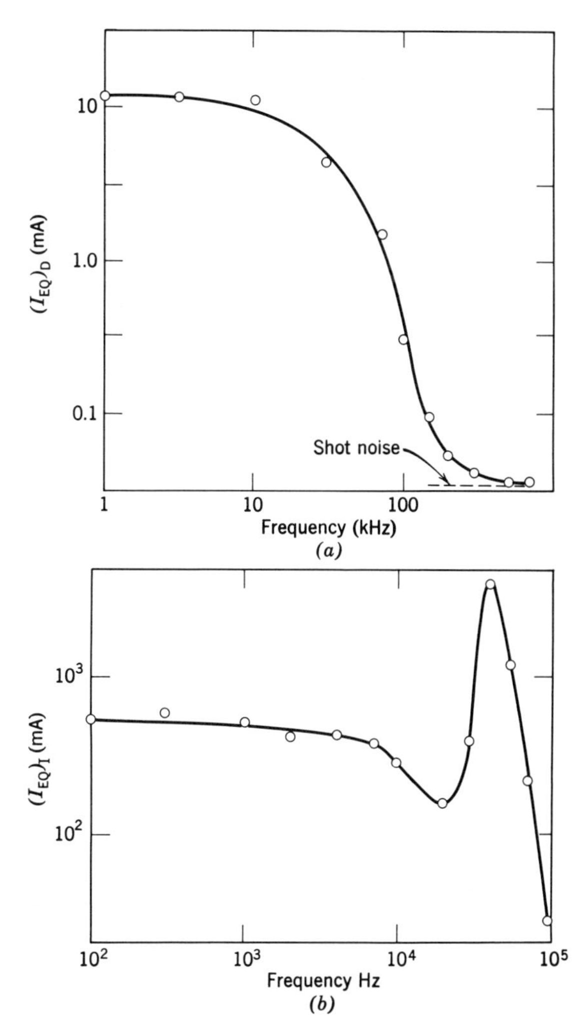

Figure 45. (*a*) Laser light noise versus frequency. (*b*) Discharge current noise versus frequency. Note that frequency scale is different. (Reproduced by permission from [92].)

by these investigators and others show that when a helium-neon laser has a noisy output of the type shown here, the noise is always coexistent with similar noise in the light output observed out of the side of the discharge tube. There will not necessarily be any time correlation between the laser noise and the side tube noise unless there is a single discrete disturbance that is propagated down the tube, such as a moving striation. Such striations are not generally observed (by eye) in helium-neon lasers unless the gas is contaminated by impurities such as oxygen or nitrogen. Nevertheless, the output of a helium-neon laser may have noise whose fluctuations are of the order of 10 or 20% of the total intensity under the worst possible conditions of high pressure and high current. In a helium-neon laser, noise can generally be suppressed by operating at low pressures, low currents, or with RF excitation. If a helium-neon laser is excited primarily by dc, addition of small amounts of RF will often reduce or eliminate the plasma noise.

The plasma fluctuations in gas lasers other than the helium-neon laser have not been studied as thoroughly. Little is known about plasma fluctuations in the CO_2 laser, for example. In the CW ion lasers certain types of lasers tend to have pressure instabilities that give rise to large-scale changes in gain and operating conditions in the laser with time constants generally of the order of 1 sec. These pressure instabilities are more noticeable in the heavier gases such as krypton and xenon than they are in argon. Dc-excited argon discharges apparently do not have large-scale sources of noise other than possible pressure instabilities, but it is not known whether the phenomena have been studied thoroughly. RF-excited ring discharges show modulation of the light at the radio frequency if the exciting frequency is of the order of 1 MHz or less, but if the frequency is raised to approximately 10 MHz the modulation becomes almost unnoticeable.

Mode interaction noise is the low frequency variation in output amplitude of a multimode laser caused by time variations in the nonlinear interactions between modes. The effect exists because, in almost all multimode lasers, no particular attention is paid to the length stability of the resonator, so that the length is generally a constantly varying function of time as the resonator varies in temperature, or because of vibrational effects. The origin of this noise can be understood qualitatively from a consideration of the hole burning effects of each mode on the population distribution under the Doppler profile.

A multimode laser will generally have many holes burned in the Doppler profile, some of which are mirror images of other holes. Thus, if the resonator changes in length, certain groups of holes will move in one direction across the Doppler profile and their images will move in

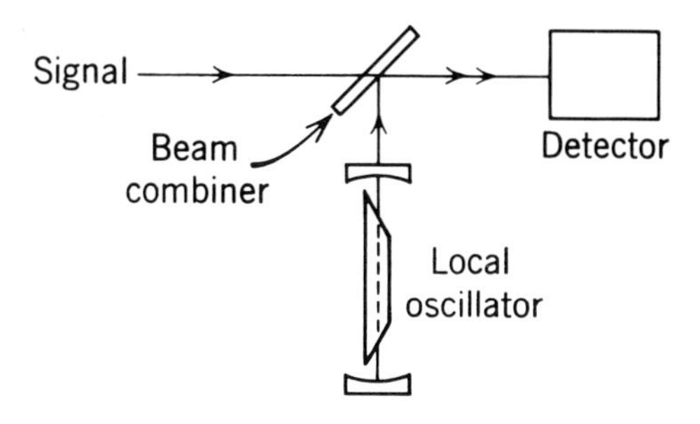

Figure 46. Elements of optical heterodyne detector.

the opposite direction. As holes coalesce and pass through each other, the nonlinear interaction defined by (30) or (33) causes changes in amplitude in the power output of the modes responsible for the holes, in a manner equivalent to the power dip in the case of a single-mode laser. Since the spacing between the modes and the various holes may be a rather complicated function, and since the time variation may also, to some extent, be random, the resultant effect is an essentially random variation in amplitude whose frequency spectrum depends in detail on the number of modes, the power output of the laser, and the rapidity of variations in the resonator length. For most lasers this type of noise is observed as approximately a 5 to 10% variation in amplitude with time constants of the order of several seconds to one minute. This source of noise can be eliminated by operating the laser in only one mode, or, as a better alternative in the case of high-power lasers, by phase locking the modes, which is discussed in the next section.

3. Detectors

Noise in the output signal caused by photodetector noise may be a problem if the light signal being detected is weak or if one is looking for very small degrees of modulation on a large input signal. For ordinary photosensitive detectors (e.g., vacuum photocells or silicon detectors) the noise contribution of the detector is the same, of course, regardless of whether the incoming light signal is laser light or incoherent light. Equation 87, which is well-known, describes the fluctuation noise of a vacuum photodetector with zero dark current; it applies to detection of laser beams, where e is the electronic charge, I the photocurrent, and $\Delta\nu$ the final receiving bandwidth:

$$I_{noise} = \sqrt{2eI\Delta\nu}. \tag{87}$$

In the detection of signals on laser beams, however, there exists a possibility for reducing the noise level that is not available when detecting incoherent light signals. This is the heterodyne detector. Use of the heterodyne detector at optical frequencies is essentially the same as at radio frequency and requires the interference mixing of a local oscillator with the desired light signal. This is shown in Figure 46. For many applications the local oscillator may be the same laser that gives rise to the signal, in which the local oscillator light beam and signal beam have previously been separated by a beam divider, and in which case the mixed signal is a homodyne signal unless the signal beam has been frequency modulated in some manner. If the signal beam and local oscillator beam come from separate lasers, heterodyne detection is practical only if both lasers are stabilized single-frequency lasers operating at very nearly the same optical frequency. The beat frequency in the combined light beam contains, as a primary carrier, the difference frequency between the two lasers. Since the optical frequency is of the order of 2×10^{14} Hz, variations in either optical frequency of the order of parts in 10^8 will shift the beat note by many megahertz. This is a factor that makes heterodyne detection between two lasers often a rather difficult experiment.

If a heterodyne detector can be used, either in a homodyne system or by making use of stabilized lasers, then it offers particular advantages in terms of the output signal-to-noise ratio. It can be shown, by an analysis which is quite similar to a corresponding analysis in the case of RF heterodyne detectors, that the use of the heterodyne detector eliminates the problem of intrinsic or dark current noise in a noisy detector and reduces the effective noise level of the detector to that given by (87), limited only by the quantum efficiency of the detector. The problem involved in designing a good heterodyne detector, aside from the local oscillator problem, is the geometrical one of insuring that the two incoming light wave fronts, when they are mixed at the detector, are essentially identical in size and in uniformity of phase across the surface. If this condition is not maintained, the heterodyne signal detected by the photodetector is degraded and so is the signal-to-noise ratio.

For a detailed discussion of laser heterodyne detectors, see Siegman [93].

D. CONTROL OF LASER COHERENCE BY INTERNAL CAVITY PERTURBATION

Up to this point we have assumed that maximum temporal coherence in a laser was obtained only by operating it under single-frequency conditions. Although single-frequency lasers do in fact have the maximum possible coherence permitted by their design, they generally have the disadvantage of operating at relatively low powers, as has been emphasized. In this section we consider various schemes that have been proposed and studied experimentally to increase the relative coherence of a high-power, naturally multimode laser by properly coupling the outputs of the modes in amplitude and phase, or by suppressing certain temporal modes altogether. Three types of outputs can be expected from lasers in which internal cavity control is employed: (*a*) true single-frequency operation, (*b*) a frequency modulated optical carrier, and (*c*) an amplitude modulated optical carrier. All of these outputs are "coherent" in the sense that the amplitude and optical frequency of the laser output at any instant of time is well defined and related to a particular carrier frequency which can be considered essentially the single frequency output. Whether or not the relative coherence of an AM or FM output is usable in a particular application depends, of course, on the application.

The various schemes to be discussed all involve resonators that are somewhat more complicated than the two mirror resonators that have been assumed up to now. The first scheme—mode selection by interferometry—requires only the addition of a third mirror and beam splitter. All other schemes involve electro-optic devices to be placed within the cavity, which entails a certain amount of additional loss as well as the complication of additional electronic equipment to drive the optically active element electronically.

1. Mode Selection by Interferometry

The general idea in this type of mode selection is to devise a resonator whose active resonator modes are determined not only by the $c/2L$ separation of the main resonator elements but by other factors such that only one mode can be sustained under the Doppler profile. The simplest scheme of this sort, and one which was experimented on at an early date, employs three mirrors in tandem, so that output can be obtained only when the stable modes of the individual resonators corresponding to each pair of widely spaced mirrors are coincident. Another system which is optically equivalent to this is shown in Figure 47. This will be referred to as the "obvious" method of mode filtering, obvious because it is equivalent optically to three resonator mirrors in tandem.

In Figure 47*b*, a beam splitter is used to split the energy leaving the laser tube more or less equally to the two end mirrors, which then reflect the energy and recombine it at the beam splitter. Such a resonator has a transmission characteristic analogous to that of a Fabry-Perot etalon with a narrow spacing: it is highly transmitting within certain narrow band passes and opaque otherwise. By making the band separations large compared to the Doppler width, one may expect to be able to select only one of numerous modes that would otherwise oscillate under the Doppler profile. Although this scheme works after a fashion in low-power, low-gain lasers, it is quite unsatisfactory in high-gain situations. In order to understand the flaw in the arguments one must consider the overall situation in the laser resonator, including the gain and undesired losses as well as the transmission of the mirror assembly.

Figure 47*c* shows the true situation when the internal losses are small compared to the gain at the peak of the Doppler profile. The vertical lines define modes which, in a two-mirror resonator, would correspond to oscillating modes because for them there is sufficient gain under the Doppler profile and they satisfy the $c/2L$ condition. The highly selective transmission of the resonator assembly is shown by the inverted curve marked "transmission" superimposed over the Doppler curve. From this

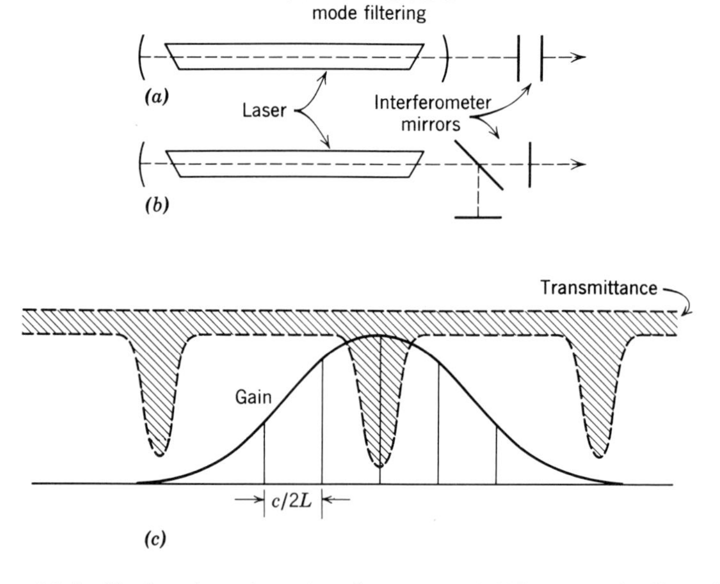

Figure 47. Mode filtering through an interferometer. (*a*) Fabry-Perot in direct line of output. (*b*) Equivalent system using two end mirrors and a beam splitter. (*c*) Transmittance superimposed on gain curve.

it can be seen that, although the transmission can be made high for one particular mode, the fact that the underlying *reflectivity* is high for all other modes tends to drive the oscillation harder in those modes. If the holes burned by the various modes become comparable to the spacing between internally oscillating modes, then this type of operation may actually draw energy away from the desired mode and place it instead into the undesired modes. Furthermore, if the peak transmission of the resonator assembly is sufficiently high to suppress oscillation, then clearly we have a situation that is exactly opposite to the one that was intended. This means that the resonator assembly, rather than selecting one out of a number of modes, suppresses a particular mode while allowing the others to oscillate unimpeded. Clearly this is not a desirable situation for obtaining single-frequency output.

The "correct" method of mode filtering was demonstrated by P. W. Smith [94] and is shown in Figure 48. Superficially it looks identical to Figure 47 except that the beam splitter is turned at 90° to its original position. In fact, however, the difference is much more subtle. The relative reflectivity of the beam splitter for transmission and for 90° reflection has to be carefully chosen, as is also the case with the reflectivities of the mirrors. The system can be analyzed by assuming that there are two resonators, consisting of mirrors M_1 and M_3, respectively, and mirrors M_4 and M_3, respectively (not mirrors M_1 and M_4, as is the case in Figure 47). The resonator consisting of mirrors M_1 and M_3 generates the $c/2L$ spacing of the primary modes of the laser resonator. The resonator M_3, M_4 determines the transmission of the assembly in the direction of the dotted line. Bear in mind that transmission in this direction occurs both by reflection of light emerging from the plasma tube and from light reflected by mirror M_4. The energy leaving in the direction of the dotted line will generally be relatively high except when interference conditions are such that the wave leaving from M_4 is 180° out of phase with that from M_1, in which case little or no energy will emerge in the direction of the arrow.

The result of this arrangement is a characteristic quite opposite to that of the Fabry-Perot resonator: the overall *transmission* of the system is very high at all frequencies except for certain narrow pass bands at which the transmission in the direction of the arrow is low and the internal reflectivity is high. The useful laser output is taken most easily out of the opposite end of the laser, from mirror M_1. The operation of the system is shown in Figure 48*b*. Note in this case that the loss curve, generated primarily by the transmission in the direction of the dotted arrow, is superimposed on the Doppler gain curve. Stable oscillation can

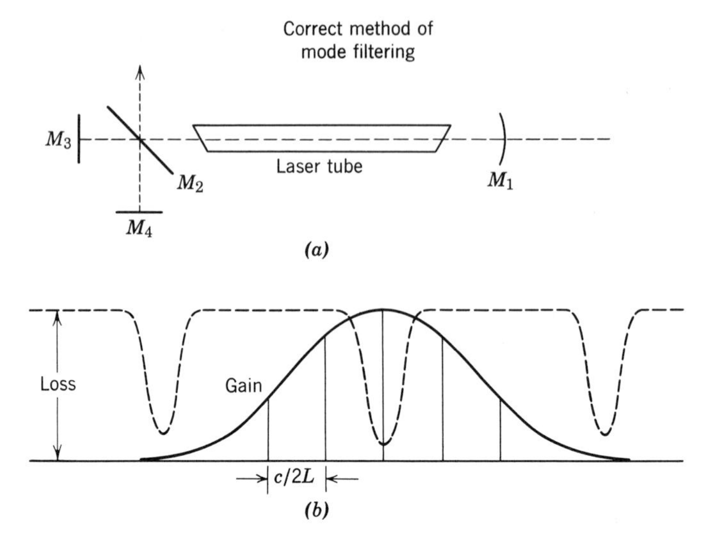

Figure 48. Laser using Smith interferometer.

occur only for those modes for which the net gain is greater than the loss caused by the selective resonator. Since this loss is reduced only in relatively narrow-band notches, laser oscillation can occur only when one of the $c/2L$ modes coincides with a notch. All other modes are suppressed by high resonator losses (not by high reflectivity, as is the case with the system shown in Figure 47). True single-frequency operation is obtained in this way.

The successful demonstration of this system by Smith [94] also showed that a substantial part of the total power contained in the total inverted population under the Doppler curve could be obtained at one frequency by this method. The success of the method depends upon properly choosing reflectivities and minimizing internal losses so that the hole burned by the single operating mode is comparable in width to the entire Doppler curve. If this is not the case, then obviously only a small fraction of the total available power will be obtained out of the laser. Smith also discussed schemes for stabilizing the length of the resonators to obtain a stabilized single frequency. The problem is clearly considerably more difficult than is the case with the short lasers discussed earlier.

Length stabilization is always a relative matter, and it is a question of stabilizing a length ratio $\Delta L/L$ where L is the total resonator length. However, the stability of the laser depends upon ΔL itself, not on the ratio. Thus as L is made larger the ratio must be made smaller for ΔL to remain the same. The resonator in question here consists of the Mirrors M_1, M_3 in Figure 48, and if this length is of the order of 1 m, for example, the stabilization problem is clearly ten times greater than it is for a 10-m-long resonator. In addition there is the requirement of controlling the length of resonator M_3, M_4 so that its notch coincides with the center of the Doppler curve. Since this resonator is quite short, the actual length stabilization is not as severe as that for the primary cavity; nevertheless, two cavities needing to be stabilized instead of one complicates the design problems.

2. AM Phase Locking

Coupling between modes, the relative amplitudes and phases being such as to produce an amplitude modulated carrier as an output, was the first type of phase locking to be demonstrated. The work of Hargrove, Fork, and Pollack [95] was performed by inserting a time-varying internal loss into the laser, and this is generally the method by which AM phase locking is obtained. However, this type of phase locking has been known to occur spontaneously in lasers without any attempt to couple

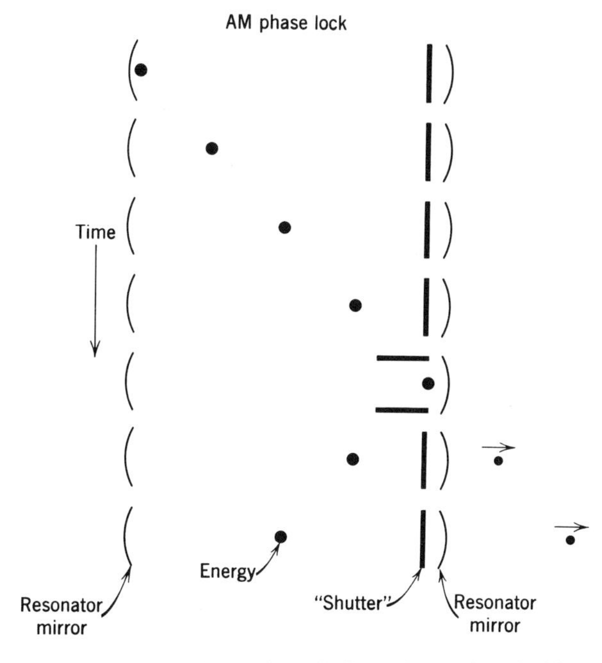

Figure 49. Pictorial description of effect of AM phase locking.

the modes together. Since the conditions under which spontaneous mode locking occurs are not clearly defined, we shall concern ourselves here only with AM mode locking generated dynamically by internal time-varying perturbations.

The general principle of AM phase locking is illustrated in Figure 49. In this illustration the internal time-varying perturbation is shown schematically as a shutter that has only two possible states, "open" and "closed." The shutter is placed as closely as possible to one of the resonator mirrors and is opened only periodically in short pulses. The pulse repetition frequency is $c/2L$, the normal frequency difference between adjacent pairs of modes. Consider a small unit of laser light energy propagating within the laser, as indicated by the large dots in the figure. If this pulse of energy is present at the mirror at the time that the shutter is open, it will be reflected from the end mirror, traverse the laser structure, gaining energy in the process, and at the end of the round trip it will arrive back at the shutter at the time that the shutter has opened again. Upon reflection, some of the energy is transmitted through the mirror, the remainder retraverses the laser and repeats the process.

It is quite obvious from this simple description that energy within the laser resonator occupying any other region of space or time will not arrive at the shutter at the time it is open and will not generate a self-sustaining wave within the laser. Thus the output of the laser consists of bursts of energy that are timed to correspond to the times that the shutter is open, that is, they have the pulse repetition frequency $c/2L$. It is also clear that if the shutter frequency is other than $c/2L$ (or an integral multiple of this quantity), it will not be possible to sustain an oscillation within the laser, since no propagating wave within the laser would be able to arrive consistently at the shutter at the time it is open.

In point of fact the actual operation of dynamic AM phase locking is essentially that of the simplified description previously given, except that it is not necessary to "close" the shutter to any great extent. All that is required generally is a periodically time-varying loss within the laser, the minimum loss being as low as possible (in keeping with general gas laser practice) and the maximum loss being only a few percent greater. The general idea here is not that it is necessary to "close" the shutter as much as it is to modulate the energy content of each mode and thus introduce sidebands at the shutter modulation frequency. When the shutter modulation frequency is $c/2L$, the sidebands of the various modes are superimposed on the unshifted carriers of these modes and thereby the internal modulation permits transfer of energy between modes. The manner in which energy is transferred between modes then depends upon the relative phases of the unshifted carriers of each mode and of the sidebands of other modes that are superimposed on it.

In AM phase locking as described here these phases are always such as to generate an amplitude modulated wave. With this type of modulation the output of the laser has its peak at the instant of minimum loss of the modulator, and the shape of the pulse is merely the Fourier transform of the modes generating the pulse with their amplitudes determined by the Doppler profile. Thus, suppose that the laser gain curve is a Gaussian and that there are 20 oscillating modes under the profile. The output pulse will correspondingly be a Gaussian whose carrier frequency will be that of the central frequency of the Gaussian gain curve (approximately) and whose width will be approximately one-twentieth of the pulse repetition rate. It is important to note that because the response of the atomic systems, which have various velocity distributions, is not particularly different whether the modes are locked or unlocked, the total output power of the laser is *approximately* the same. Thus for the example given, the peak power in each pulse of the phase locked output will be approximately 20 times greater than the output that could be obtained with an unlocked laser.

Experimentally, several different means may be used to provide the time-varying internal loss. The method used in the original experiment of Hargrove, Fork, and Pollack [95] employed an ultrasonic wave in a quartz transducer, such that a periodic standing wave of density was set up in the quartz. When the wave intensity reached its peak, the quartz was strained to produce a diffraction grating that was capable of diverting energy from the straight-through beam into auxiliary beams that would miss the laser mirrors and thus divert energy out of the resonator. The driving frequency for such a transducer is one-half the $c/2L$ frequency, since the standing wave disappears and reverses phase (thus generating maximum transparency) at every half cycle.

More commonly, the transducer is an electro-optic modulator employing a material such as KDP or lithium niobate, in which a shift in output polarization is achieved by applying an electric field across the material. In this particular case, the exact details of the polarization shift—to elliptical or to some other type of polarization—are not important provided that they provide a component perpendicular to that which is transmitted most easily by the Brewster windows. Thus when the polarization of the wave passing through the crystal is shifted, part of the energy is diverted by reflection off the Brewster window and the desired effect is obtained. With this system, as with the quartz ultrasonic transducer, the driving frequency is at the half frequency, since presumably the system has maximum transmission at the instant that the driving field is zero. When electro-optic crystals are placed inside a resonator cavity, special care must be taken to keep the losses as low as possible, particularly those caused by protective windows and other boundary surfaces, which often cannot be placed conveniently at Brewster's angle. Special design and the use of particularly good antireflection coatings on surfaces are often necessary in this application.

For a considerably more detailed theoretical treatment of phase locking than the one just sketched, the reader is referred to a paper by Di Domenico [96]. The original experimental paper [95] and other papers that have been published since give dramatic proof of the effectiveness of AM phase locking in their illustrations of the observed optical output spectrum. In an AM phase locked laser, all oscillating modes are present with exactly the $c/2L$ spacing and with amplitudes determined by the Doppler profile, and these amplitudes remain constant as a function of time. In an unlocked laser, on the other hand, the spacing is often somewhat uneven owing to nonlinear interactions between modes, and the amplitudes of the various modes wander, largely as a function of time. This phenomenon is, of course, the mode interaction noise discussed earlier, which is completely absent in an AM phase locked laser. An incidental bonus of AM phase locking is the fact that the average power output of an AM phase locked laser is often several percent greater than the average power of the same laser when not phase locked, presumably because of a more uniform saturation of the atomic system.

3. FM Mode Coupling

It is also possible to couple modes in such a way that the output represents a frequency modulated signal. The spectrum of a frequency modulated signal constituting a low-frequency sinusoidal FM of a carrier, with no accompanying amplitude modulation, is given by the following well-known formula:

$$\begin{aligned} F(t) &= \cos(\omega + \Delta \cos \Omega t) \\ &= J_0(\Delta) \cos \omega t - J_1(\Delta)[\sin(\omega+\Omega)t + \sin(\omega-\Omega)t] \\ &\qquad - J_2(\Delta)[\cos(\omega+2\Omega)t + \cos(\omega-2\Omega)t] \\ &\qquad + J_3(\Delta)[\sin(\omega+3\Omega)t + \sin(\omega-3\Omega)t] \\ &\qquad + \ldots, \end{aligned} \tag{88}$$

where the J's are Bessel functions. The coefficients of the spectral terms at radial frequencies $\omega + n\Omega$ differ from those obtained in the case of amplitude modulation in the following ways: (a) the phase relationships are quite different, since in the FM case each sideband is 90° out of phase with adjacent sidebands; and (b) under certain conditions the carrier or any of the sidebands may have zero amplitude. Because of these properties, which do not exist in the AM case where all spectral components are present and in phase, it is not obvious a priori that one can induce FM phase locking unless one can "order" one of the frequency components to be the carrier and the others to be the sidebands. Harris, in his review paper [97], points out that if a choice of carrier can be made, it will be indefinitely sustained as a stable oscillation condition. If the gain properties of the laser medium were completely linear, there would be an arbitrary choice of which of the locked temporal modes became the carrier, but because of the saturation properties of the laser medium, it generally occurs that the mode closest to the peak of the Doppler curve becomes the carrier automatically with all others serving as sidebands.

An example of an oscillation condition corresponding to FM mode coupling is shown in Figure 50, taken from a paper by Ammann, McMurtry, and Oshman [98]. It can be seen in FM mode coupling that the intensities of the various modes do have the type of distribution to be expected from (88). Since the heights of the components shown in Figure 50 do not correspond to a Doppler profile, it is not clear that the full potential power output of the laser medium can be obtained in this way; however, if hole burning is extensive, the loss in potential output power is not likely to be very large.

In the case of AM phase locking the phase locking was induced by imposing a variable loss modulation on the laser resonator. In order to obtain FM mode locking one would therefore suspect intuitively that the experimental procedure to be used would be that of introducing a modulated phase—one modulating the mode spacing itself—at the $c/2L$ frequency. Such modulation of the cavity length would be done most effectively by physically moving one of the mirror surfaces, but it has not been feasible up to now to move such a massive object as a mirror at frequencies of the order of 100 MHz. The next best technique, which is generally used, is to modulate the equivalent optical path length within the resonator by employing an electro-optic element, properly oriented, across which an oscillating electric field is placed. It turns out, furthermore, that in order to obtain FM mode locking, the driving frequency for mode coupling must be detuned somewhat from the $c/2L$ frequency; in fact, if the modulation is attempted exactly at $c/2L$, one gets AM mode locking instead of FM locking. An explanation of this phenomenon is complicated and will not be attempted here. The reader is referred to the review paper by Harris [97] and to earlier source papers on the topic for a mathematical discussion of this phenomenon as well as the ranges of frequency detuning over which one gets AM or FM mode locking.

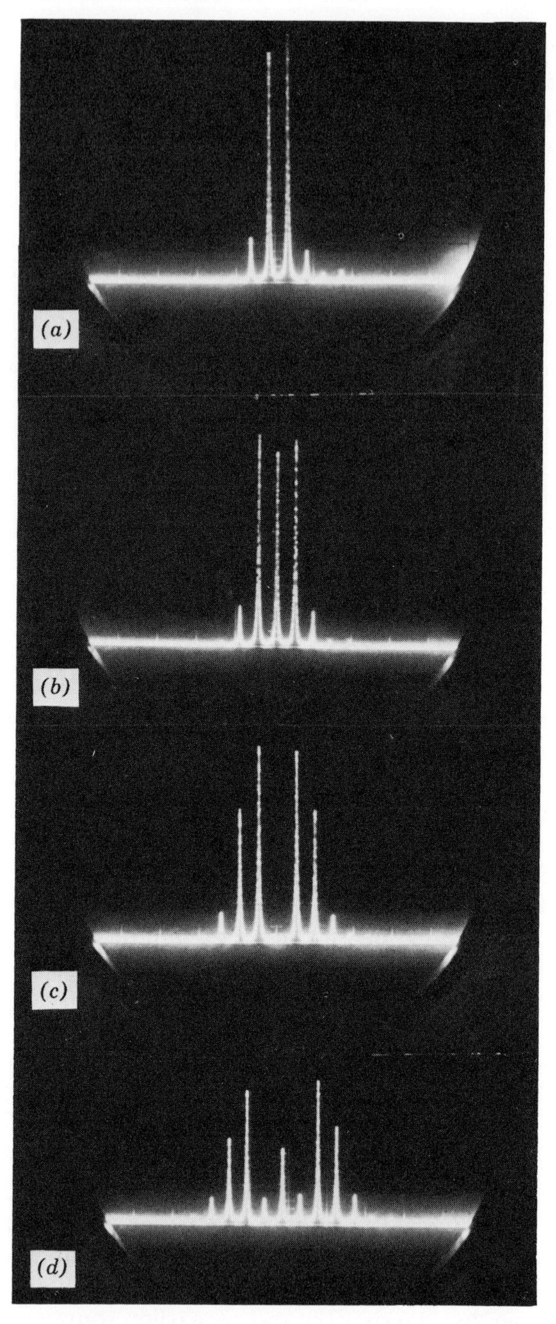

Figure 50. Optical spectra obtained with a high-resolution interferometer showing relative power in the temporal modes. (*a*) Free-running. (*b*), (*c*), and (*d*) Conditions of FM mode coupling. (Reproduced by permission from [98].)

As in the case of AM mode locking, it is possible in FM locking to lock at multiples of the $c/2L$ frequency; it is also possible to lock groups of modes independently in such a manner that the output consists of two or more essentially independent frequency modulated signals. Although such conditions are of mathematical interest, the situation that one usually strives for in generating mode locking, either FM or AM, is that of locking all modes together to produce a single modulated carrier as the output.

4. Application of Internal Perturbation Methods to High-Power Single-Frequency Lasers

One of the most important applications of the internal perturbation techniques that we have been discussing is that of developing a single-spatial-mode, single-temporal-mode laser output that has the high power typical of a large multimode laser. Of the various schemes that have been proposed for this purpose, three appear to be the most potentially useful. All three have been demonstrated in the laboratory, but each scheme has certain advantages and disadvantages for possible commercial exploitation in the future. The three single frequency schemes are shown schematically in Figures 51 and 52.

Figure 51 shows the first two schemes, of which (a), called "correct" mode filtering, has already been discussed. It makes use of two resonators, one short and one long, to determine the single-frequency output, and has the advantage that no active electronic elements are needed within the cavity. The disadvantages of the scheme are, first, that two resonator cavities have to be stabilized instead of one, and, second, that

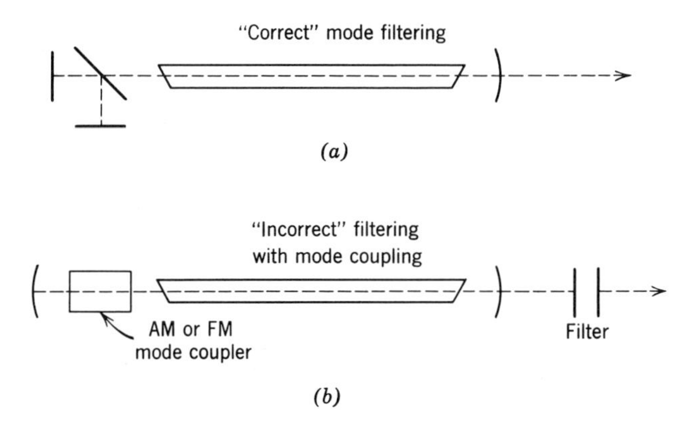

Figure 51. Two single-frequency schemes. (*a*) Smith interferometer laser. (*b*) "Incorrect" filtering with mode coupling.

hole burning will not saturate the entire laser medium unless the laser intrinsically has high gain and high power output. Thus the scheme would not be very practical for use with a laser having intermediate gain and power.

The second scheme makes use of "incorrect" filtering of the type shown in Figure 47, but with the difference that an AM or FM mode coupling device is inserted in the resonator. In Figure 51*b*, the mode filter is shown as a straight-through Fabry-Perot rather than as an interferometer, as in Figure 47, however, the effect is the same. With mode coupling, the major disadvantage of the "incorrect" filtering method—most of the laser energy is "bottled up" in unwanted modes—is removed because the effect of active mode coupling is to allow energy to be transferred freely back and forth between modes within the laser. Thus, if the Fabry-Perot filter is chosen to have an optimum transmission for the desired mode, a steady state situation can be reached in which almost all of the energy is transferred out of the laser in that particular mode; the energy remaining in the other modes is merely equivalent to stored energy in the resonator which is available to the output coupler. This may be considered the analog, in the frequency domain, to the following situation in the space domain: In a multimirror resonator, outputs can be obtained from any one of the mirrors, however, only one of them is normally chosen to be the transmitting mirror. The output of this mirror is chosen to be an optimum and the outputs of the others are suppressed insofar as possible. Going back to the single-frequency scheme, then, the Fabry-Perot filter should be designed to have maximum opacity for the unwanted modes and an optimum transmittance for the desired mode. This optimum transmittance is not the optimum that would be calculated using simple formulas such as (38), but it must be based on the rate of energy transfer between modes and the total energy available within the system. The reader is referred to Harris [97] and references listed therein for the mathematical details of calculating this optimum transmittance.

This particular single-frequency scheme has the advantages of moderate simplicity and noncritical requirements with regard to the mode coupler, since it may be either AM or FM. Furthermore, with proper transmission out of the Fabry-Perot, essentially all of the available energy within the laser is available in the single output mode. A disadvantage of the system is that since the Fabry-Perot cannot have complete opacity for all of the unwanted modes, some energy will leak out of the resonator from these unwanted modes; it will be a much smaller proportion of the total than would be the case with incorrect filtering without mode coupling; nevertheless, these components will be present and may cause some difficulties in applications.

The third scheme, shown in Figure 52, has been given the name of the "supermode" scheme. The principle of this scheme is based on the fact that a given frequency modulated signal can be considered equivalent to a phase modulated signal, with the proper choice of frequency or phase modulation coefficients. If we thus consider the output of an FM mode coupled laser to be a phase modulated wave, then a single-frequency output may be obtained by demodulating the phase variations in the laser output. In Figure 52, the same RF oscillator, operating in the vicinity of the $c/2L$ frequency, drives both the electro-optic modulator within the laser and the electro-optic demodulator outside the laser. The demodulator is of identical construction to the modulator, although with considerably greater optical length, and is phased to cancel the phase variations in the laser output. Thus when the output of the laser is at the high-frequency end of its frequency excursion, corresponding to "compression" of the waves of the output carrier, the demodulator should be "expanding" the wave, and vice versa.

A problem in the use of this scheme is the relatively long optical path length required in the demodulator, particularly when the number of modes to be coupled is large and the Doppler width is also relatively large, as is the case with the argon ion laser. The output of a properly mode coupled FM laser has excursions of the instantaneous frequency that cover approximately the entire range of frequencies of the various modes that make up the modulated wave. For a typical multimode laser, the frequency excursions in the laser output thus cover approximately the Doppler width. On the other hand, the frequency excursions caused

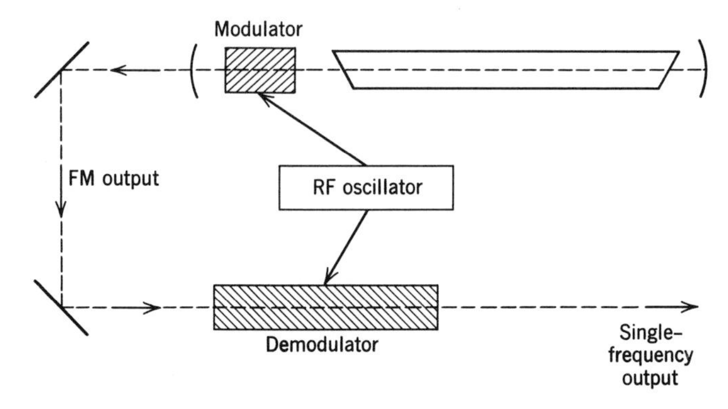

Figure 52. "Supermode" scheme for obtaining single-frequency output at high power.

by the phase modulation of the modulator itself, required to perform the mode coupling operation, certainly need not be more than the $c/2L$ frequency difference, and mode coupling can generally be obtained when the frequency excursions within the modulator are only a small fraction of this number. Thus, whereas the modulator need be only a small electro-optic element, perhaps a few millimeters of material such as potassium dihydrogen phosphate (KDP), the demodulator may need as much as 1 m of path length under some extreme conditions. This poses a severe materials problem. Some advantage can be gained by folding the optical path within the demodulator by multiple reflections, and some further advantage can be obtained by employing materials that are particularly suitable for this purpose such as lithium niobate. Even so, if a material such as lithium niobate is used to demodulate the output from an argon laser, for example, the optical path length within the demodulator will have to be at least several centimeters. Another disadvantage of the scheme is the complexity of the electronics, inasmuch as two electro-optic elements must be driven. On the other hand, they are driven at exactly the same frequency and therefore by the same RF master oscillator, and the relative phases, although dependent upon both electrical and optical path lengths in the system, are fixed for any one system.

Once these disadvantages are surmounted, the supermode scheme has a number of important advantages that may make it a particularly useful device, not only where single-frequency operation is desired at high power, but also for high stability lasers. The frequency of the single-frequency output is determined by the "carrier" mode within the laser, which is the mode closest to the center of the Doppler profile. If the cavity length should shift as a function of time, the output frequency will shift correspondingly until one of the sidebands has shifted to a position closer to Doppler center than the carrier, at which point the frequency will jump discontinuously by an amount $c/2L$ to make the new central mode the carrier. Thus the output frequency can never be more than $c/4L$ from the Doppler center, even in the absence of resonator stabilization, and if short-term variations can be ignored, there is actually an advantage in this system to making the cavity as long as possible.

For many applications where short-term stability is not important but long-term stability is important, the supermode scheme allows one to dispense with resonator stabilization altogether. On the other hand, if cavity length stabilization is still necessary, the supermode method provides a powerful means for obtaining a signal for stabilizing the cavity length with very high signal-to-noise ratio. For this purpose, a small amount of output power is taken out of the laser beam before the electro-optic demodulator (or perhaps taken out of the opposite end of the laser) and detected with a high-speed photocell. It has been shown that if the oscillating modes within the laser are symmetrically located relative to the Doppler profile, the output contains no amplitude variations whatsoever. However, if the resonator length changes so that the modes are slightly displaced from symmetry, then there are amplitude variations at the even harmonics of the $c/2L$ modulation frequency; in particular, there is a component at the second harmonic. An error signal for stabilizing the cavity length can thus be obtained by detecting variations in the laser beam output at the second harmonic frequency and phase-detecting these with the corresponding second harmonic output obtained by multiplication from the RF oscillator. The scheme is discussed in detail by Harris [97] and by Harris and McDuff [99]. The method is particularly sensitive because, first, the signal-to-noise ratio for photo-detection is taken from a high-power laser beam and is thus very high, and, second, information relative to line center is obtained by "sampling" the entire Doppler profile, thus locating very accurately a point of symmetry, rather than by sampling only one or two points. The method can also be made quite insensitive to short-term amplitude variations due to plasma noise, should this be necessary.

At the present time none of the high-power single-frequency schemes discussed here are available to users of commercial lasers, nor does it appear likely that any will be available in the near future. It is also not entirely clear at present which of the three methods will ultimately be the most widely used.

CHAPTER 5

PROBLEMS ASSOCIATED WITH LASER APPLICATIONS

The applications of gas lasers have become so varied that it is not practical to attempt to discuss more than a small fraction of them. In this chapter brief mention is made of the most common and important applications, and some discussion is given of problems that arise in connection with applying lasers to these applications.

A. LONG-DISTANCE METROLOGY

The use of the finite velocity of light for distance measurement has been recognized for some time. With lasers, as with incoherent sources, two different regimes may be recognized. For short distances, interferometric methods may be used, and distance measurement then becomes a matter of counting fringes or wavelengths. For longer distances, this method is not practical; thus one proceeds to chop the light beam into segments and measure distance by the travel time for each segment. When a laser is used as the light source, particularly a single-frequency laser with a long coherence length, the dividing line between short-distance interferometry and long-distance time-of-flight measurement is determined not by the visibility of fringes as much as by refractive index variations in the air which might wash out fringes. In typical atmospheres this dividing point may be of the order of several meters, although individual laboratories have performed interferometry measurements in air over distances as long as 100 m. In this section we consider the problems involved in long-distance measurement, where the distances may be of the order of many kilometers.

Distance measurement by the time-of-flight method has been performed using many different types of lasers, both pulsed and CW, gaseous or otherwise. A typical system design for use with a CW gas laser is shown in Figure 53. In this system the CW laser output (which need not be single frequency) is amplitude modulated by an electro-optic modulator at frequencies generally of the order of 30 to 50 MHz. The light is beamed at the distant target and return radiation is collected through a telescope and detected by a photocell whose output then contains a major component at the modulation frequency. The photocell output is phase detected and the final output is placed into the measurement or recording device as a signal directly proportional to phase delay. The distance to the target is determined as the phase delay, expressed in distance, plus N times the wavelength of the modulation frequency, where N may be a large unknown number. A variety of techniques can be used to determine N. One possibility is that additional modulation signals at lower frequencies can be imposed on the laser output and each of these modulation frequencies can be separated in the photocell output and processed separately, which eventually reduces the ambiguity to that of the lowest modulation frequency, for example, 1 MHz. Eventually a point is reached at which visual observation or previous knowledge of rough distance eliminates any remaining ambiguity. Basically, the important feature of laser beam long-

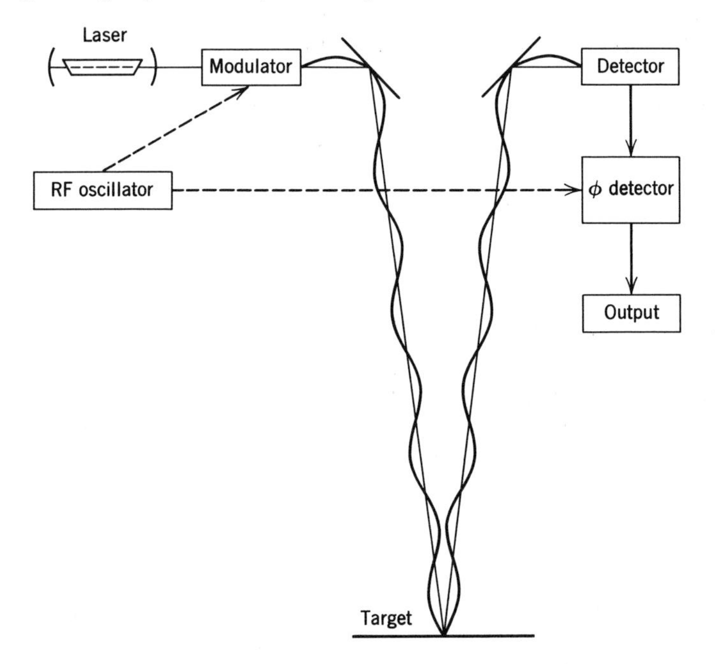

Figure 53. Use of a CW laser and modulator for distance measurement.

distance measurement is the high differential accuracy that it provides for measuring small displacements at long distances, and for this purpose it is the highest modulation frequency that provides the greatest usefulness, since the accuracy is determined only by the final uncertainty in phase, and the corresponding distance uncertainty is then inversely proportional to frequency.

The method described in Figure 53 is not basically different from that employing incoherent light sources, which has been used with success for a number of years. One may then ask what advantages are gained by employing a laser instead of an incoherent source as the light source. First, because of the very narrow beam diameter and high collimation of a laser output, the electro-optic modulator can be made very small, and this makes it possible to use much higher modulation frequencies than would otherwise be possible. Second, the collimation allows one to concentrate the beam on a target having a very small area, thus allowing one to study high spatial resolution at long distances and maximizing the signal-to-noise ratio. Finally, the high monochromaticity of a laser beam output allows the use of very narrow band filters at a photocell to remove unwanted radiation such as sunlight. This is extremely important if one wishes to use the distance measurement technique in the daytime.

The accuracy of the distance measurement is generally determined either by signal-to-noise ratio in the output of the photocell, or in many cases where this is large, by the electronic uncertainties of measuring relative phases in the phase detector. When a 10-mW laser beam is employed as the source, signal-to-noise ratio becomes a problem only when the target is a thoroughly diffused, dark object at a distance of about 10 km and the experiment is carried out in broad daylight. The exact signal-to-noise ratio of such an operation in daylight depends on the width of the interference filter in the optical detection system, which at the present time is generally of the order of 0.1 Å. If a "cooperative" target, for example, a retro reflector, is used as the target, the distance at which highly accurate phase measurements can be obtained is limited only by the visibility and may be as much as hundreds of kilometers. If such measurements were carried out in a vacuum, they would allow the absolute distance to be specified to within the phase uncertainty, which at a 50-Mc modulation frequency would be of the order of a few millimeters. In air, a considerable additional uncertainty is introduced because of unknown density variations in the air and the need to correct for its refractive index. Owens [100] has written a review of these problems and has given a particularly accurate formula for the group velocity (i.e., the group refractive index) of air under various conditions. Usually, over very long distances the density of the air over many parts of the optical path is not known to sufficient accuracy. For this purpose, it would be desirable to perform at the measurement at two or more widely separated wavelengths, so that the mean density itself can be calculated.

B. METROLOGY BY INTERFEROMETRY

One of the most common applications of gas lasers is to interferometry and this, in turn, is generally employed for one of two purposes: either for testing of optical surfaces, which will be discussed, or for very accurate distance measurement making use of the known wavelength of the light source. We discuss here two questions of importance regarding the use of single-frequency lasers in short-distance metrology.

1. Reproductibility of Present Wavelength Standards

Since interferometers have been devised which can measure to small fractions of a fringe, indeed to thousandths of a fringe, the accuracy of such a measurement is ultimately determined by the accuracy to which the laser wavelength is known in absolute units. It is fortunate at least that the international primary standard of length—the krypton orange line at 6057 Å—is also an optical wavelength in the visible, so that interferometric measurements may be used to compare any laser directly with the international primary standard. Such measurements have been carried out in the standards laboratories of several countries, generally with the helium-neon laser at 6328 Å. The present best accepted value, for an He-^{20}Ne laser operating at the center of the power dip, appears to be $\lambda = 6329.9138$ Å in a vacuum [101, 102]. A slightly higher value [103] published recently by the United States National Bureau of Standards is being rechecked, and it is expected that a revised value will be available soon from that laboratory. Based on this value for the vacuum wavelength, Appendix B gives tables of the wavelength in air for various atmospheric conditions.

It is not yet known in great detail to what extent the output wavelength is likely to change between one stabilized single-frequency laser and the next. Possible sources of shifts in a single-frequency helium-neon laser are factors that may shift the center of the Doppler profile itself, such as pressure-dependent shifts in both the helium and the neon components, and shifts in the center of the power dip relative to the Doppler profile, if use is made of the power dip for stabilization purposes. Experiments [74] on 6328-Å lasers centered on the minimum of the power dip show variations as a function of both pressure and gain, of the order of parts in 10^8 (approximately 4 MHz) at pressures of the order of 1 torr, and as much as one part in 10^7 (30 to 40 MHz) at pressures above 3 torr (see Figure 54). Although these may be considered

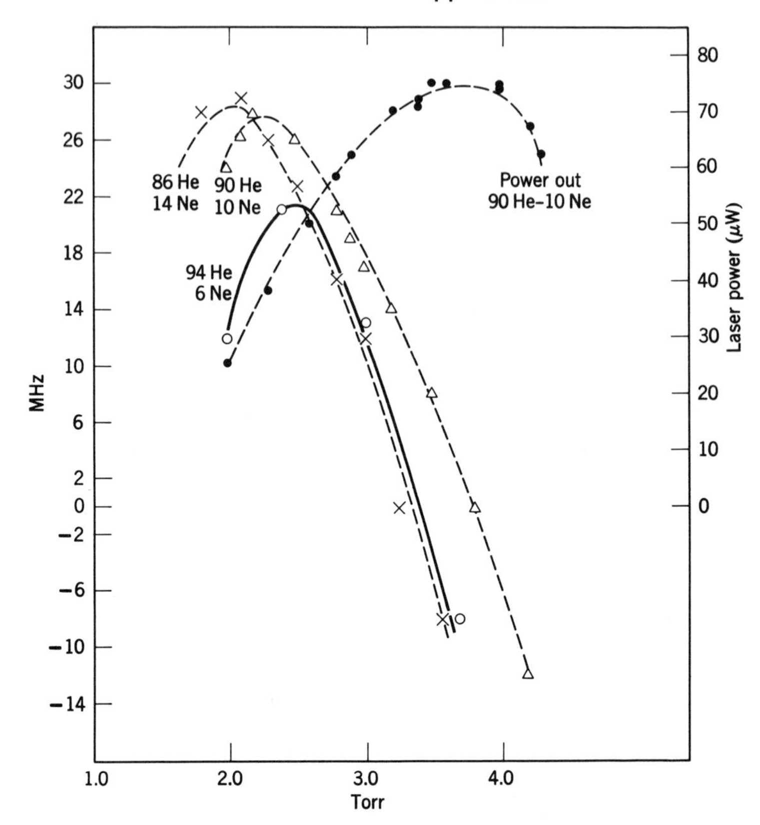

Figure 54. Experimental pressure shifts in a 6328 Å laser stabilized on the power dip. The numbers in front of the He and Ne symbols refer to percentages in the mixture. (From [74].)

the possible extremes of variation, the numbers must be taken to be the relative uncertainty of the output wavelengths of any given laser because, although the pressure of the plasma tube when new may be known to approximately 0.1 torr, this pressure may change over the operating life of the laser tube, and the gain-to-loss ratio is a relatively uncontrollable factor unless Brewster windows and mirrors are kept clean at all times.

Similar experiments on the power dip at 1.15 μ performed by intercomparing two lasers show effects similar to those at 6328 Å but considerably smaller in percentage [73]. This is to be expected because the source of the pressure shift arises largely from the atomic interactions of atoms in the upper laser state during collisions and near-collisions.

Moreover, the 3s level, which gives rise to 6328 Å, is at higher energy and involves electrons that are more loosely bound than those of the 2s level, which gives rise to the 1-μ radiation, so that 6328 Å would be expected to have greater sensitivity to pressure effects.

Experiments have also been performed on pressure shifts of a laser stabilized by a passive resonant element of the type shown in Figure 37, and these experiments have demonstrated that, since the output wavelength depends on the Gaussian profile without a power dip, the pressure shift effects are somewhat less [77]. In fact, it is possible to choose a composition (relative mixture and pressure) of the gas in the passive reference cell that minimizes pressure and current dependent shifts and may make possible even higher accuracy in the wavelength determination for this type of laser. This is a particular advantage of the system described in Figure 37 for metrology, since the composition of the reference cell can be chosen to optimize the stability of the reference wavelength, whereas in a laser plasma tube the optimization of gain is the overriding factor. Other lasers with single-frequency output, for example, the supermode type, may be expected to have different dependences upon gas pressure, current density, and gain. The supermode laser, not depending on a power dip but on symmetry of the Gaussian, may be expected to have smaller shifts than those of lasers stabilized on the power dip, but the gas mixture cannot be controlled for minimum pressure shift as is the case with the passive cell reference system.

2. Problems in the Coupling of Interferometers to Lasers

One of the most severe problems that the user of a single-frequency laser encounters in applying his laser to metrology is the interaction of the laser with the measuring device. The problem is exactly analogous to that of interaction of a microwave oscillator with reflected energy from an improperly matched microwave device, and the methods of solution are also quite analogous. In both cases, the problem disappears if no energy is returned back to the source, but impedance matching in optics may be a much more difficult problem than it is in microwaves.

The nature of the problem, for a particularly simple example of a laser coupled to a Fabry-Perot interferometer, is shown in Figure 55. The problem is not significantly different for coupling to other types of interferometers. The figure shows a laser, presumably of single frequency and stabilized to some reference point on the Gaussian profile, whose output passes through one lens or more and then, through the interferometer, to the detector. Each surface in the lens system, the interferometer system, and the detector is capable of reflecting energy back in the general direction of the laser. As the interferometer is changed in length, the detector counts "fringes," alternate variations of

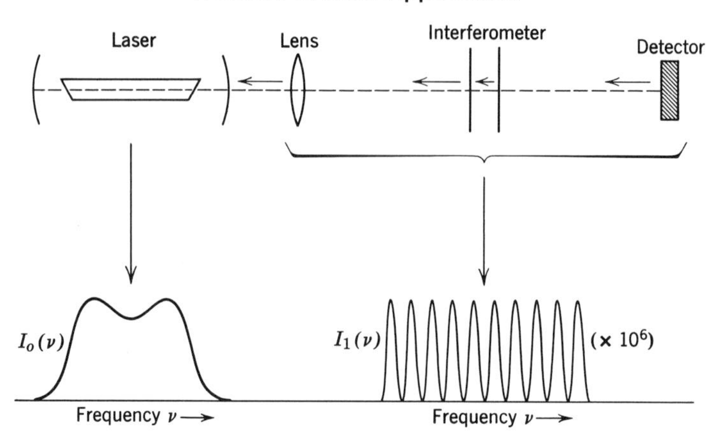

Figure 55. Coupling of laser and interferometer.

high and low light intensity. When the intensity at the detector is low, the interferometer is returning maximum energy in the direction of the laser, and for most interferometers the optimum adjustment would be that which would return the energy exactly in the direction of the laser beam. This is a highly undesirable situation, but fortunately most interferometers will work under just slightly less than optimum conditions by a slight misadjustment of the perpendicular of the interferometer so that the main reflection does not impinge directly upon the laser output mirror. Similarly, lens surfaces may be given an antireflection coating (which in practice will never be perfectly antireflecting) and also adjusted to avoid direct back-reflections. Nevertheless, all known solid optical surfaces have small-angle scattering, which, although not large, may be sufficient to scatter enough energy back into the laser to cause difficulties, and the detector surface itself is also capable of returning small amounts of energy to the laser by diffused reflection. An estimate of the amount of returned energy that might cause uncontrollable frequency shifts will now be given.

Consider the output characteristic of the laser itself in Figure 55. Let this be defined as $I_0(\nu)$. Consider now a secondary resonator structure consisting of one of the laser mirrors (the back mirror) and one or more of the diffused reflecting surfaces of the external optical system. This weak "interferometer" produces a change in saturated gain of the laser such as to modify the output characteristic slightly from I_0, so that it can now be described by

$$I(\nu)=I_0(\nu)+I_1(\nu), \tag{89}$$

where I_1 is the contribution back into the laser from the external interferometer. Now the laser is stabilized at the point at which $d^2I/d\nu^2 = 0$ so that, if I_1 is present, the operating frequency is shifted from the point $d^2I_0/d\nu^2 = 0$ by approximately

$$\Delta\nu = -\frac{dI_1/d\nu}{d^2 I_0/d\nu^2}. \tag{90}$$

For a helium-neon laser, $d^2I_0/d\nu^2$ is of the order of $I_0 \times 10^{-18}$ Hz^{-2}, a number that depends primarily on the Doppler width and is essentially independent of the stabilization scheme. A low-finesse interferometer of any form has fringes that are approximately sinusoidal in nature, and thus I_1 has the form

$$I_1 = I_0\left[A + B\cos\left(\nu_0 + \frac{\nu L}{\pi c}\right)\right], \tag{91}$$

where A and B are constants including attenuation and L is the length of the secondary "interferometer." If L is in centimeters, $dI_1/d\nu$ is approximately $I_0 \times 10^{-11}$ BL in maximum absolute value. Inserting this and the numerical value for $d^2I_0/d\nu^2$ in (90), we obtain the result that a shift $|\Delta\nu| = 1$ MHz can be obtained if $BL \approx 0.1$. The actual shift can lie anywhere between $+\Delta\nu$ and $-\Delta\nu$ and is likely to fluctuate rapidly between these values since the secondary interferometer is not particularly stable. To introduce specific values, with an interferometer 1 m in length, frequency fluctuations of the order of 1 MHz would result from a value of B of 0.1%. This is a typical length for an optical system attached to a single-frequency laser, and the small amount of reflected energy required to cause shifts indicates directly the severity of the problem.

Control of this problem requires meticulous care on the part of the experimenter to reduce scattered light into the laser to as small a value as possible. On some occasions, experimenters can make use of "one-way" devices such as circular polarizers and optical Faraday rotators to reduce back-reflected energy, but caution must be used that the front surfaces of these devices do not in themselves reflect back more energy than might be returned from the optical system.

C. DATA PROCESSING AND HOLOGRAPHY

These applications make use of the transform properties of light as it is diffracted and propagated through space. For these applications, in general, it is essential that the wave front be a pure sphere or plane without phase reversals in the electric field at any point, that is, corresponding to the lowest-order spatial mode. It is often desirable, also, that there be no amplitude variations in the wave front, a condition that cannot be satisfied directly from the laser output since, as we know, this generally has a Gaussian intensity variation as a function of distance from the axis of propagation. For most work in holography, the Gaussian intensity distribution is not a serious problem, but in optical data processing, where transforms must be taken of detailed intensity variations after the light passes through a mask, the Gaussian intensity variation may cause problems in determining the true transform of the mask. Various schemes have been proposed to make the amplitude distribution more uniform, including special optical systems and photographically produced masks. A photographically produced mask is simply a negative produced by exposing a piece of fine grain film in the laser beam, at the position at which it is to be used, and developing the film to a gamma of one. Such a procedure, of course, entails a resulting loss in total intensity available for use in the data processor.

The development of the art of holography has arisen almost entirely, except for original proposals, since the advent of the gas laser and is largely dependent on the laser for present and future developments. Since there are numerous excellent review articles in the literature regarding the fundamental principles of holography, and since a detailed discussion would take us far afield of the present topic of gas lasers, we shall not take it up here. The following remarks concern the properties of the lasers and of the laser beam that is to be used in holography.

A typical setup for making a hologram, using a gas laser, is shown in Figure 56. The laser light output provides both the reference beam, which constitutes the greater part of the light that exposes the hologram plate, and also the subject illumination, whose scattered light interferes with the reference beam at the site of the hologram plate. The two condi-

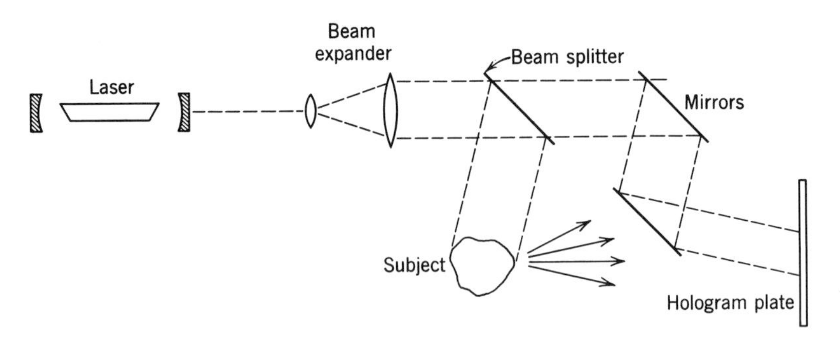

Figure 56. Method of hologram production.

tions that are essential for hologram production are: (a) the path difference between the two optical paths (reference beam and scattered light from subject) must be within the coherence length of the laser, and (b) the subject must not move optically more than one-tenth wavelength during exposure, or the interference fringes that constitute the hologram will be smeared out. The second requirement suggests that an ideal light source for making holograms should be a pulsed laser having a very short pulse of sufficient intensity to expose the plate at once. Such holograms have been made, using *Q*-switched ruby lasers, but pulsed lasers capable of supplying sufficient power in one pulse have so far been difficult to combine with the requirement for sufficient coherence. The coherence requirement limits the front-to-back distance of the subject, unless special steps are taken to illuminate different parts of the object with different optical paths. The "coherence length' of the laser, determined by the number of operating temporal modes, determines the depth of the subject. For multimode lasers, therefore, the following limitations apply roughly to the depth of the object that can be photographed as a hologram: helium-neon, approximately 30 cm; argon or krypton ion lasers, approximately 5 cm. Of course, if single-frequency lasers of sufficient power can be used, then the coherence length becomes much longer and this limitation on subject depth disappears. In such cases the only limitation on subject depth would be obtaining sufficient intensity to make exposures in reasonably short times. Under present multimode conditions, holograms of typical tabletop subjects (the size of a telephone, for example) can be made in a few seconds using a 10 mW helium-neon laser.

Another requirement of the laser used in hologram production is that the beam be "clean," that is, free of extraneous scattered light. This condition is usually obtained by aperturing in the optical system that follows the laser.

The gas laser has also turned out to be a convenient light source for reconstructing the hologram image, although the requirements on the laser for this purpose are not so stringent as they are for hologram production. High spatial coherence (the lowest-order mode) is desirable because each illuminated point on the reconstructed image is essentially also a diffracted image of the illuminating source. Thus a multispatial mode or high-order mode laser, appearing as a relatively diffused source, will wash out details in the reconstructed hologram image. The coherence required to determine maximum path length difference, unlike the interference requirement in making the hologram, is determined by diffraction grating equations, since this is effectively the way a hologram is used in making the reconstructed image. If a portion of the hologram of width D is used for reconstructing a given part of the reconstructed image, and the hologram is viewed at an angle θ from the normal to the hologram plate, then the path length difference for which it is required that the light be coherent is $D \sin \theta$.

For any gas laser, even an ion laser having only a few centimeters coherence length, problems of temporal coherence in reconstruction would arise only if a large hologram, say, 15 to 20 cm long, were used in its entirety to reconstruct an image in great detail. Normally, if a hologram is viewed by eye, the portion of a hologram that is subtended by the eye in defining any given point in the reconstructed image is determined by the pupil of the eye, having the diameter of several millimeters, and any gas laser then makes a satisfactory source. It is, of course, essential that the laser be operating on only one laser wavelength, since different wavelengths will give rise to reconstructed images of slightly different apparent sizes and locations.

D. OPTICAL TESTING

One of the most common uses of the gas laser is that of a laboratory tool; for example, to test optical systems, as a light source for interferometry of optical surfaces, and as a teaching aid to demonstrate the laws of physical optics. For these purposes, the attributes of the gas laser are its small effective spot size and relative monochromaticity, as well as its portability and compactness. High output power is generally not important in these applications, and temporal coherence enters only if interferometry is to be performed with relatively long path differences. However, even a simply constructed multimode laser is considerably more coherent than ordinary light sources and this has made possible several types of interferometry for use in optical testing that are not generally possible with incoherent sources, particularly tilted plate interferometry. Figures 57 and 58 illustrate some typical uses of the laser in an optical laboratory.

Figure 57 shows a laser used for aligning an interferometer, use being made of the fact that the back reflections of the laser spot are very sharply defined and can be made coincident upon the front face of the laser to a high degree of accuracy. In the figure surfaces 1, 2, and 3 are adjusted in turn so that the reflected light of the laser beam is directed onto, and partly through, the aperture. When all three reflected images have been superimposed, the interferometer is automatically in alignment.

Figure 58 shows the laser being used for surface parallelism testing. This is a very common use of the gas laser which measures relative paral-

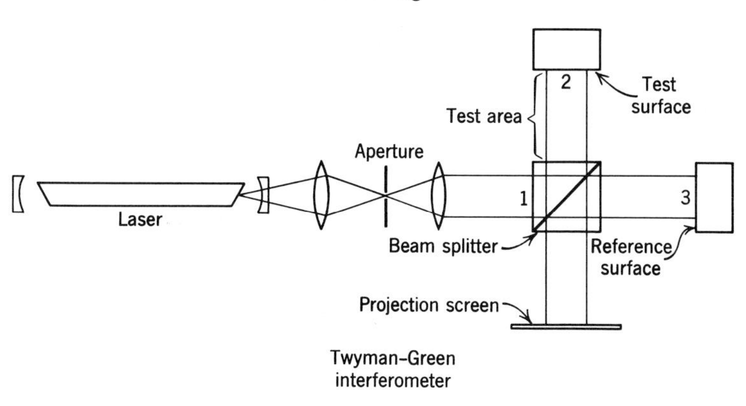

Figure 57. Use of laser to align an interferometer.

lelism of two surfaces. It is particularly useful for checking optical quality of a flat window. The coherence requirement on the laser is only that the path difference between beams reflected from the front and back surfaces be within the coherence length of the laser, a condition that is usually quite easily satisfied. The fringes that are observed on the projection screen are always sharp and very well defined, since uncoated flats will have reflections of equal field amplitudes from the two surfaces so that the dark parts of the fringes are truly at zero intensity.

E. SCATTERING, INCLUDING RAMAN AND BRILLOUIN SCATTERING

The properties of the gas laser that make it useful for studying Raman effect and associated effects are the relative monochromaticity and high degree of collimation. Generally it is not necessary to have the laser either single spatial mode or single frequency, although there are specialized versions of these applications in which such properties may be desirable. For studying Raman effect in liquids, the high collimation is

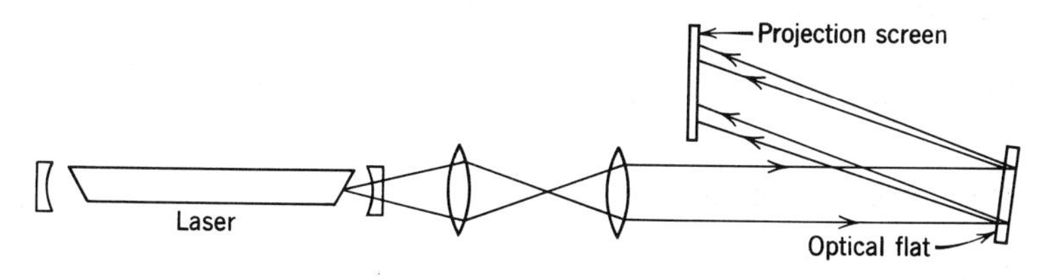

Figure 58. Laser test of surface parallelism.

the most important property, because it makes possible experiments with a high degree of discrimination between the Raman scattered light and the incident light. Other properties of the laser beam that are useful for this purpose are the high degree of linear polarization that is obtainable from a Brewster window laser, and the ability to observe Raman scattering over a very wide range of angles relative to the incident beam. These two conditions make possible an analysis of the Raman effect in tensor form where this is applicable from scattering in solids, and this provides much more information regarding chemical structure than can be obtained from previous types of Raman scattering experiments.

Although Raman signals are typically very weak, the signal-to-noise ratio that is possible with a properly constructed Raman spectrometer is so great that lasers of only moderate power are required. It appears that 100 mW in any given wavelength is generally sufficient to satisfy the experimenter who wishes to observe even the weakest Raman signals. The helium-neon laser at 6328 Å has been extensively used for this purpose because, at the longer wavelength, Rayleigh scattering is greatly minimized. When such experiments are to be performed with blue or green light, ionized argon lasers are quite adequate.

For certain experiments involving studies of Brillouin and Rayleigh scattering, the frequency shifts of the scattered light relative to the incident light may be so small they are beyond the resolution of the spectrometer. In such cases heterodyne detection systems have proved useful. If a single-frequency laser can be used, then, of course, there is no ambiguity regarding the relative frequencies of the reference signal (the unscattered light from the laser) and the light scattered off the sample. However, there are many situations in which multifrequency lasers can be used just as easily. This is true if the frequency shift of the scattered light is either small compared to the mode frequency spacings or relatively large compared to them. In either case there is a heterodyne beat note between each frequency in the incident beam and its corresponding shifted frequency; however, these difference frequencies are all essentially the same so that they are simply combined in the heterodyne detector and appear as one beat signal. Thus the use of multifrequency lasers for these studies is quite common, since such lasers usually have considerably more output power than can be obtained easily with single-frequency lasers.

F. HIGH ENERGY DENSITY

The use of high energy density to burn holes in materials and to obtain nonlinear optical effects has, until recently, been almost entirely the province of the pulsed solid-state laser. However, the output of a high-power ionized argon laser or of a CO_2 laser now makes burning experiments quite feasible. Many nonlinear optical effects can be studied in this way also, particularly by the use of pulsed gas lasers having higher peak powers, or by the use of Q switching in a CO_2 laser. However, many of the interesting nonlinear optical effects and multiphoton processes that require megawatt peak powers are still beyond range of available gas lasers.

A discussion of the theoretical aspects of obtaining high energy density from a gas laser is essentially that of the optical system that focuses the beam to a small spot. The reader is referred to Chapter 3, Section A2 of this text where the question of focusing at small F numbers has already been taken up.

G. SAFETY PROBLEMS ASSOCIATED WITH WORKING WITH GAS LASERS

There are two types of safety problem that enter in working with gas lasers. One is the electrical problem generated by the fact that high voltages are present in any gas laser in order to maintain the electric discharge. This problem is not particularly different from the electrical safety problems attendant upon any other instrument employing high voltages. The other problem is the optical problem based on the fact that the laser delivers a concentrated beam of high-intensity radiation. This is a relatively new problem that has arisen only because of the advent of the laser, and many aspects of it have not been studied. Furthermore, most of the studies of the interaction between laser radiation and living tissue that have been performed to date have involved pulsed high-intensity lasers such as the ruby laser. It thus appears that there is very little knowledge about the detailed interaction of CW gas laser radiation with tissue, and when safety standards are promulgated for such lasers it is generally with a very generous safety factor, since the threshold for harmful effects is often not known to within several orders of magnitude of the intensity of the incident radiation.

Damage to surface tissue such as the skin or the cornea of the eye arises from heating. However, the effects of radiation at different wavelengths are different, depending on the absorption of the tissues. Wavelengths in the red and near infrared appear to penetrate skin quite easily, whereas blue and green are absorbed more rapidly. Radiation at 10 μ is absorbed in a very thin layer (of the order of several microns thick) because of the complete opacity of water to this radiation. The sensitivity of different tissues to heating varies also. For example, the tissues that make up the cornea and crystalline lens of the eye are particularly sensitive to small rises in temperature and must be particularly shielded from any radiation that may be absorbed in these tissues to a degree sufficient to cause a rise in temperature.

Damage to the retina can occur if visible or near infrared radiation is focused by the eye to a small spot on the retina. The effect here is, again, generation of high temperatures in the tissue, although in this case the temperature rise will occur in a very small region corresponding to the size of the focused spot. It is in regard to the threshold for retinal damage that there appears to be the least amount of agreement among qualified specialists in the field. Some laboratory experiments have been performed on laboratory animals in which the eye has been immobilized and exposures of specified lengths of time have been given through specially designed focusing optics. The experimenter or bystander in a laboratory where visible lasers are employed is not likely to be in situations even approaching the ideal for damage conditions. He will be moving about, his eyes will be moving about rapidly; if laser light enters it will probably not be focused on the retina in perfect focus, and in most cases (unfortunately not all) his immediate response will be to blink and to look away from the laser beam.

Safety standards, unfortunately, cannot be based on the assumption that this situation will always be the case. The image of the sun on the retina is much more intense than the image of most laser beams that are likely to enter the eye, yet most people walk around in bright sunlight and do not appear to suffer retinal damage because of it. On the other hand, it is known that retinal damage has occurred under conditions where people have stared directly at the sun without adequate protection. One should therefore not discount the effects of curiosity when laser beams are around in places where they may directly enter the human eye.

The following remarks are intended to establish minimum safety standards in laboratories or under field conditions where lasers are used, and they appear to agree with remarks made by qualified specialists which have been published at various times. The reader, however, is reminded that I have had no training in ophthalmology and have not had these suggestions reviewed by a qualified ophthalmologist.

The standards are based on the assumption that CW gas laser beams have diameters typical of the mode diameters of a laser beam without additional optics, that is, visible lasers whose output diameter is of the order of 1 to 3 mm and CO_2 lasers with diameters of the order of 1 cm.

1. Visible Laser Intensities of One to One Hundred Microwatts

At the 1 μW level it is generally agreed that the laser beam can enter the eye directly and be studied (i.e., stared at for an indefinite length of time) without fear of damage. At somewhat higher intensities there is disagreement about the threshold for damage. The shorter the viewing time, clearly the less chance of damage up to viewing times of the order of 3 sec. A good rule of thumb might be this: if there is discomfort, or if afterimages persist for more than a few seconds, it is probably better to reduce the intensity by means of neutral density filters.

2. One to One Hundred Milliwatts

The laser beam, focused and unfocused, may be safely displayed on a white diffuse reflecting surface and viewed as any other object. If the beam is focused to a point, the rule about afterimages should apply. Precautions should be taken that such a laser beam cannot enter the eye directly; however, if it does so and if the eye is moved out of the beam quickly, the probability for damage is minimized. (Many laboratory benches and laser setups are at a height corresponding to the eyes of children. It is a good idea to permanently and rigidly prohibit children from visiting any laboratory where lasers are used at bench height.)

3. One to Ten Watts, Visible, Near Infrared, or CO_2 Laser Radiation

Special precautions should be taken that the unattenuated laser beam can never enter the eye under any circumstances. At an intensity of 1 W a visible laser beam 1 mm in diameter impinging on a white diffuse reflecting surface will generate a surface intensity comparable to that of the surface of the sun. If the surface is not a diffuse reflector, the intensity when viewed in certain directions may be considerably stronger than that of the sun. For this reason it is a good idea, when working with such laser beams, either to spread the beam out to diameters of 1 cm or larger with beam expanding optics, or to prohibit the viewing of such beams on light-colored surfaces. Beams of this intensity, when incident on skin, can usually be felt very strongly; damage will be kept minimal if the beam is treated in the same manner as any other hot object.

4. One-Hundred Watt Beams, Ion Lasers, and CO_2 Lasers

At this intensity level, there is a great danger of injury, either to the eye or to any other tissue, if the beam hits the tissue even for an instant. Special precautions must be taken that the beam is at all times in a closed space and can never be scattered in a specular manner out of the enclosure for any reason. Accidents with high-power CO_2 lasers have been known to occur when a test object in the beam cracked and moved, causing a momentary specular reflection to flash across the skin of a nearby experimenter. Safety glasses should be worn at all times in a laboratory in which such lasers are in operation. If there is any possibility at all of specular reflection from a 100-W CO_2 laser escaping into the room, the safety glasses should be heavy transparent plastic or the type coated with a thin metallic reflecting surface, not merely dark glasses. A 100-W CO_2 laser beam impinging on ordinary glass or thin plastic safety lenses will crack them and possibly shatter them completely, so that they provide no safety protection at all.

If pulsed gas lasers, rather than CW lasers, are used, the preceding remarks will need some modification, although it is not certain to what extent the damage thresholds may be lowered. There is evidence that some of the retinal damage effects that have been observed with lasers are dependent on peak power, rather than average power, in which case the safety thresholds for pulsed lasers would have to be reduced by the amount of the duty cycle of the laser. On the other hand, if a laser is operating in many spatial modes, then the image of the laser on the retina of the eye will be considerably more diffuse than would be the case for a single-mode laser and it may be possible to relax the restrictions somewhat. In this case, the rule regarding afterimages is a good one, provided that one starts with a very high density filter as an attenuator and, if necessary, gradually reduces the attenuation.

BIBLIOGRAPHY

[1] A. L. Bloom, "Gas Lasers," *Appl. Opt.,* **5**, 1500-1514, and *Proc. IEEE,* **54**, 1262-1276 (October 1966).

[2] W. V. Smith and P. P. Sorokin, *The Laser,* McGraw-Hill, New York, 1966.

[3] B. Lengyel, *Introduction to Laser Physics,* Wiley, New York, 1966.

[4] G. Birnbaum, *Optical Masers,* Suppl. 2, *Advan. Electron. Electron Phys.,* Academic, New York, 1964.

[5] F. Bloch, "Nuclear Induction," *Phys. Rev.,* **70**, 460 (1946); see also N. Bloembergen, *Nuclear Magnetic Relaxation,* Benjamin, New York, 1961, Pp. 21-49.

[6] N. Bloembergen, *Nonlinear Optics,* Benjamin, New York, 1965, Pp. 20-61.

[7] W. E. Lamb, Jr., "Theory of an Optical Maser," *Phys. Rev.,* **134**, A1429-A1450 (June 15, 1964).

[8] W. R. Bennett, Jr., "Gaseous Optical Masers," *Appl. Opt.* (Suppl. 1, *Optical Masers),* Pp. 24-61 (1962).

[9] J. P. Goldsborough, E. B. Hodges, and W. E. Bell, "RF Induction Excitation of CW Visible Laser Transitions in Ionized Gases," *Appl. Phys. Lett.,* **8,** 137-139 (March 1966).

[10] R. L. Byer, W. E. Bell, E. Hodges, and A. L. Bloom, "Laser Emission in Ionized Mercury: Isotope Shift, Line Width, and Precise Wavelength," *J. Opt. Soc. Am.,* **55**, 1598-1602 (December 1965).

[11] D. A. Leonard, "Saturation of the Molecular Nitrogen Second Positive Laser Transition," *Appl. Phys. Lett.,* **7**, 4-6 (July 1965).

[12] A. L. Schawlow and C. H. Townes, "Infrared and Optical Masers," *Phys. Rev.,* **112**, 1940-1949 (1958).

[13] A. Javan, W. R. Bennett, Jr., and D. R. Herriott, "Population Inversion and Continuous Optical Maser Oscillation in a Gas Discharge Containing a Helium-Neon Mixture," *Phys. Rev. Lett.,* **6**, 106-110 (February 1961).

[14] C. K. N. Patel, "Vibration Energy Transfer—An Efficient Means of Selective Excitation in Molecules," in *Physics of Quantum Electronics,* P. L. Kelley, B. Lax and P. E. Tannenwald, eds. McGraw-Hill, New York, 1966, Pp. 643-654.

[15] A. G. Fox and T. Li, "Resonant Modes in a Maser Interferometer," *Bell Syst. Tech. J.,* **40**, 453-488 (1961).

[16] G. D. Boyd and J. P. Gordon, "Confocal Multi-Mode Resonator for Millimeter through Optical Wavelength Masers," *Bell Syst. Tech. J.,* **40**, 489-508 (March 1961).

[17] G. D. Boyd and H. Kogelnik, "Generalized Confocal Resonator Theory," *Bell Syst. Tech. J.,* **41**, 1347-1369 (July 1962).

[18] A. D. White and J. D. Rigden, "Continuous Gas Maser Operation in the Visible," *Proc. IRE* (Correspondence), **50**, 1697 (July 1962).
[19] L. E. S. Mathias and J. T. Parker, "Visible Laser Transitions from Carbon Monoxide," *Phys. Lett.*, **7**, 194-196 (November 1963).
[20] W. E. Bell, "Visible Laser Transitions in Hg^+," *Appl. Phys. Lett.*, **4**, 34-35 (January 1964).
[21] W. B. Bridges, "Laser Oscillation in Singly Ionized Argon in the Visible Spectrum," *Appl. Phys. Lett.*, **4**, 128-130 (April 1964); see also G. Convert, M. Armand, and P. Martinot-Lagarde, "Transitions Lasers Visibles dans l'Argon Ionisé," *Comptes Rendus*, **258**, 4467-4469 (1964); and W. R. Bennett, Jr., J. W. Knutson, Jr., G. N. Mercer, and J. L. Detch, "Superradiance, Excitation Mechanisms, and Quasi-CW Oscillation in the Visible Ar^+ Laser," *Appl. Phys. Lett.*, **4**, 180-182 (May 1964).
[22] E. I. Gordon, E. F. Labuda, and W. B. Bridges, "Continuous Visible Laser Action in Singly Ionized Argon, Krypton, and Xenon," *Appl. Phys. Lett.*, **4**, 178-180 (May 1964).
[23] C. E. Moore-Sitterly, *Atomic Energy Levels*, Vol. I, II, and III. National Bureau of Standards Circular 467 (U. S. Govt. Printing Office).
[24] W. R. Bennett, Jr., E. A. Ballik, and G. N. Mercer, "Spontaneous-Emission Line Shape of Ion Laser Transitions," *Phys. Rev. Lett.*, **16**, 603-605 (April 1966), and earlier work referred to therein.
[25] M. Born and E. Wolf, *Principles of Optics*, Third Ed., Pergamon, New York, 1966.
[26] W. R. Bennett, Jr., "Inversion Mechanisms in Gas Lasers," *Appl. Opt.* (Suppl. 2, *Chemical Lasers*), Pp. 3-33 (1965).
[27] W. Heitler, *The Quantum Theory of Radiation*, Third Ed., Oxford, New York, 1954, Pp. 175-179.
[28] For some historical background, see J. Agassi, "The Kirchhoff-Planck Radiation Law," *Science*, **156**, 30 et seq. (April 7, 1967).
[29] A. C. G. Mitchell and M. W. Zemansky, *Resonance Radiation and Excited Atoms*, Cambridge, New York, 1934 (also in paperback, 1961). Pp. 92-96.
[30] D. F. Hotz, "Gain Narrowing in a Laser Amplifier," *Appl. Opt.*, **4**, 527-530 (May 1965).
[31] A. G. Fox and P. W. Smith, "Mode-Locked Laser and the 180° Pulse," *Phys. Rev. Lett.*, **18**, 826-828 (May 1967).
[32] S. L. McCall and E. L. Hahn, "Self-Induced Transparency by Pulsed Coherent Light," *Phys. Rev. Lett.*, **18**, 908-911 (May 1967).
[33] C. L. Tang and H. Statz, "Optical Analog of the Transient Nutation effect," *Appl. Phys. Lett.* **10**, 145-147 (March 1967).
[34] A. Szöke and A. Javan, "Isotope Shift and Saturation Behavior of the 1.15μ Transition of Neon," *Phys. Rev. Lett.*, **10**, 512-524 (June 1963).
[35] H. C. Torrey, "Transient Mutations in Nuclear Magnetic Resonance," *Phys. Rev.*, **76**, 1059 (1949).
[36] E. L. Hahn, "Spin Echoes," *Phys. Rev.*, **80**, 580 (1950).
[37] W. W. Rigrod, "Saturation Effects in High-Gain Lasers," *J. Appl. Phys.*, **36**, 2487-2490 (August 1965); and, "Gain Saturation and Output Power of Optical Masers," **34**, 2602-2609 (September 1963).
[38] R. H. Dicke, "Coherence in Spontaneous Radiation Processes," *Phys. Rev.*, **93**, 99-100 (January 1954).
[39] P. W. Smith, "The Output Power of a 6328 Å He-Ne Gas Laser," *IEEE J. Quant. Elect.*, **QE-2**, 62-68 (March 1966).
[40] D. C. Sinclair, "Polarization Characteristics of an Ionized-Gas Laser in a Magnetic Field," *J. Opt. Soc. Am.*, **56**, 1727-1731 (December 1966).

[41] E. I. Gordon, E. F. Labuda, R. C. Miller, and C. E. Webb, "Excitation Mechanisms of the Argon-Ion Laser," in *Physics of Quantum Electronics*, P. L. Kelley, B. Lax and P. E. Tannenwald, eds. McGraw-Hill, New York, 1966. Pp. 664-673.

[42] S. Kobayashi, T. Izawa, K. Kawamura, and M. Kamiyama, "Characteristics of a Pulsed Ar II Ion Laser Using the External Spark Gap," *IEEE J. Quant. Elect.*, **QE-2,** 699-700 (October 1966).

[43] C. K. N. Patel, W. L. Faust, R. A. McFarlane, and C. G. B. Garrett, "CW Optical Maser Action up to 133μ (0.133 mm) in Neon Discharges," *Proc. IEEE*, **52,** 713 (June 1964).

[44] H. Steffen et al., *Phys. Lett.*, **21,** 425-426 (June 1966).

[45] P. W. Smith, "On the Optimum Geometry of a 6328 Å Laser Oscillator," *IEEE J. Quant. Elect.*, **QE-2,** 77-79 (April 1966).

[46] R. N. Zitter, "$2s$-$2p$ and $3p$-$2s$ Transitions of Neon in a Laser Ten Meters Long," *J. Appl. Phys.*, **35,** 3070–3071 (October 1964).

[47] H. A. H. Boot, D. M. Clunie, and R. S. A. Thorn, "Pulsed Laser Operation in a High Pressure Helium-Neon Mixture," *Nature*, **198,** 773-774 (1963).

[48] A. L. Bloom, R. L. Byer, and W. E. Bell, "Emission Line Widths of Ion Lasers," in *Physics of Quantum Electronics*, P. L. Kelley, B. Lax and P. E. Tannenwald, eds. McGraw-Hill, New York, 1966. Pp. 688-689, also unpublished work.

[49] E. F. Labuda and E. I. Gordon, "Microwave Determination of Average Electron Energy and Density in He-Ne Discharges," *J. Appl. Phys.*, **35,** 1647-1648 (May 1964), and references contained therein.

[50] A. D. White and J. D. Rigden, "The Effect of Super-Radiance at 3.39 μ on the Visible Transitions in the He-Ne Maser," *Appl. Phys. Lett.*, **2,** 211-212 (June 1963).

[51] C. B. Moore, "Gas-Laser Frequency Selection by Molecular Absorption," *Appl. Opt.*, **4,** 252-253 (February 1965).

[52] A. L. Bloom, "Observation of New Visible Gas Laser Transitions by Removal of Dominance," *Appl. Phys. Lett.*, **2,** 101-102 (March 1963).

[53] A. D. White, "Reflecting Prisms for Dispersive Optical Maser Cavities," *Appl. Opt.*, **3,** 431-432 (March 1964).

[54] W. E. Bell and A. L. Bloom, "Zeeman Effect at 3.39 μ in a Helium-Neon Laser," *Appl. Opt.*, **3,** 413-415 (March 1964), and papers referred to therein.

[55] E. A. Ballik, W. R. Bennett, Jr., and G. N. Mercer, "Temperatures, Lorentzian Widths, and Drift Velocities in the Argon-Ion Laser," *Appl. Phys. Lett.*, **8,** 214-216 (April 1966).

[56] W. B. Bridges and A. N. Chester, "Visible and UV Laser Oscillation at 118 Wavelengths in Ionized Neon and Other Gases," *Appl. Opt.*, **4,** 573-580 (May 1965).

[57] W. B. Bridges and A. N. Chester, "Spectroscopy of Ion Lasers," *IEEE J. Quant. Elect.*, **QE-1,** 66–84 (May 1965).

[58] E. F. Labuda and A. M. Johnson, "Threshold Properties of Continuous Duty Rare Gas Ion Laser Transitions," *IEEE J. Quant. Elect.*, **QE-2,** 700-701 (October 1966).

[59] E. F. Labuda, E. I. Gordon, and R. C. Miller, "Continuous-Duty Argon Ion Lasers," *IEEE J. Quant. Elect.*, **QE-1,** 273-279 (September 1965).

[60] J. P. Goldsborough, "Cyclotron Resonance Excitation of Gas-Ion Laser Transitions," *Appl. Phys. Lett.*, **8,** 218-219 (May 1966).

[61] C. K. N. Patel, "CW High-Power N_2-CO_2 Laser," *Appl. Phys. Lett.*, **7,** 15-17 (July 1965), and earlier papers referred to therein.

[62] G. Moeller and J. Dane Rigden, "High-Power Laser Action in CO_2-He Mixtures," *Appl. Phys. Lett.*, **7,** 274-276 (November 1965); see also C. K. N. Patel, P. K. Tien, and J. H. McFee, "CW High-Power CO_2-N_2 Helium Laser," *Appl. Phys. Lett.*, **7,** 290-292 (December 1965).

[63] P. K. Cheo, "Effects of CO_2, He, and N_2 on the Lifetimes of the 00°1 and 10°0 CO_2 Laser Levels and on Pulsed Gain at 10.6 μ" *J. Appl. Phys.*, ***38,*** 3563-3568 (August 1967).

[64] J. D. Rigden and G. Moeller, "Recent Developments in CO_2 Lasers," *IEEE J. Quant. Elect.*, **QE-2,** 365-368 (September 1966).

[65] N. Wiener, *The Fourier Integral and Certain of its Applications,* Cambridge, New York, 1933, and Dover, New York, 1966. Pp. 46-71.

[66] R. Courant and D. Hilbert, *Methods of Mathematical Physics,* Vol. I, Interscience, New York, 1953. Pp. 91-97.

[67] A. Erdelyi, ed., *Higher Transcendental Functions,* Vol. 2, McGraw-Hill, New York, 1953. Pp. 188-196.

[68] H. Kogelnik and T. Li, "Laser Beams and Resonators," *Appl. Opt.*, **5,** 1550-1567, and *Proc. IEEE,* **54,** 1312-1329 (October 1966).

[69] A. G. Fox and T. Li, "Effect of Gain Saturation on the Oscillating Modes of Optical Masers," *IEEE J. Quant. Elect.*, **QE-2,** 774-783 (December 1966).

[70] T. S. Jaseja, A. Javan, and C. H. Townes, "Frequency Stability of Helium-Neon Masers and Measurements of Length," *Phys. Rev. Lett.*, **10,** 165 (March 1963).

[71] B. L. Gyorffy and W. E. Lamb, Jr., "Pressure Effects in the Output of a Gas Laser," in *Physics of Quantum Electronics,* P. L. Kelley, B. Lax and P. E. Tannenwald, eds. McGraw-Hill, New York, 1966. Pp. 602-610.

[72] A. D. White, "Frequency Stabilization of Gas Lasers," *IEEE J. Quan. Elect.*, **QE-1**, 349-357 (November 1965).

[73] A. Szöke and A. Javan, "Effects of Collisions on Saturation Behavior of the 1.15 μ Transition of Neon Studied with Helium-Neon Laser," *Phys. Rev.*, **145**, 137-147 (May 1966).

[74] A. L. Bloom and D. L. Wright, "Pressure Shifts in a Stabilized Single-Wavelength He-Ne Laser," *Appl. Opt.*, **5**, 1528-1532; and *Proc. IEEE,* **54**, 1290-1294 (October 1966).

[75] W. R. Bennett, Jr., "Mode Pulling in Gas Lasers," in P. Grivet and N. Bloembergen, eds., *Quantum Electronics III,* Columbia University Press, New York 1964. P. 441.

[76] W. R. Bennett, Jr., S. F. Jacobs, J. T. Latourette, and P. Rabinowitz, "Dispersion Characteristics and Frequency Stabilization of a Gas Laser," *Appl. Phys. Lett.*, **5**, 56 (August 1964).

[77] A. D. White, "Pressure- and Current-Dependent Shifts in the Center Frequency of the Doppler-Broadened 6328 Å ^{20}Ne Transition," *Appl. Phys. Lett.*, **10**, 24-26 (January 1967).

[78] R. L. Fork, W. J. Tomlinson, and L. J. Heilos, "Hysteresis in an He-Ne Laser," *Appl. Phys. Lett.*, **8**, 162-163 (April 1966).

[79] T. G. Polanyi, M. L. Skolnick, and I. Tobias, "Frequency Stabilization of a Gas Laser," *IEEE J. Quant. Elect.*, **QE-2**, 178-179 (July 1966).

[80] W. W. Rigrod, "The Optical Ring Resonator", *Bell Syst. Tech. J.*, **44**, 907-916 (May-June 1965).

[81] C. V. Heer, "Resonant Frequencies of an Electromagnetic Cavity in an Accelerated System of Reference," *Phys. Rev.*, **134**, A799-A804 (May 1964).

[82] W. M. Macek and D. T. M. Davis, Jr., "Rotation Rate Sensing with Traveling-Wave Ring Lasers," *Appl. Phys. Lett.*, **2**, 67-68 (February 1963).

[83] P. K. Cheo and C. V. Heer, "Beat Frequency Between Two Traveling Waves in a Fabry-Perot Square Cavity," *Appl. Opt.*, **3**, 788-789 (June 1964).

[84] S. A. Collins, Jr., "Analysis of Optical Resonators Involving Focusing Elements," *Appl. Opt.*, **3**, 1263-1275 (November 1964).

[85] B. Richards and E. Wolf, "Electromagnetic Diffraction in Optical Systems II. Structure of the Image Field in an Aplanatic System," *Proc. Royal Soc. A*, **253**, 358-379 (1959).
[86] A. Boivin and E. Wolf, "Electromagnetic Field in the Neighborhood of the Focus of a Coherent Bcam," *Phys. Rev.*, **138**, B1561-B1565 (June 1965).
[87] A. L. Bloom and D. J. Innes, *Design of Optical Systems for Use with Laser Beams*, Spectra-Physics Laser Tech. Bull. No. 5 (Spectra-Physics, Inc., Mountain View, California).
[88] E. I. Gordon, "Optical Maser Oscillations and Noise," *Bell Syst. Tech. J.*, **43**, 507-539 (January 1964).
[89] C. Freed and H. A. Haus, "Photoelectron Statistics Produced by a Laser Operating below and above the Threshold of Oscillation," *IEEE J. Quant. Elect.*, **QE-2**. 190–195 (August 1966).
[90] A. W. Smith and J. A. Armstrong, "Laser Photon Counting Distributions Near Threshold," *Phys. Rev. Lett.*, **16**, 1169-1172 (June 1966).
[91] U. Ingard, "Acoustic Wave Generation and Amplification in a Plasma," *Phys. Rev.*, **145**, 41-46 (May 1966).
[92] L. J. Prescott and A. van der Ziel, "Gas Discharge Modulation Noise in He-Ne Lasers," *IEEE J. Quant. Elect.*, **QE-2**, 173-177 (July 1966).
[93] A. E. Siegman, "The Antenna Properties of Optical Heterodyne Receivers," *Appl. Opt.*, **5**, 1588-1594; and *Proc. IEEE*, **54**, 1350-1356 (October 1966).
[94] P. W. Smith, "Stabilized, Single-Frequency Output from a Long Laser Cavity," *IEEE J. Quant. Elect.*, **QE-1**, 343-348 (November 1965).
[95] L. E. Hargrove, R. L. Fork, and M. A. Pollack, "Locking of He-Ne Laser Modes Induced by Synchronous Intracavity Modulation," *Appl. Phys. Lett.*, **5**, 4-5 (July 1964).
[96] M. DiDomenico, Jr., "Small-Signal Analysis of Internal (Coupling Type) Modulation of Lasers," *J. Appl. Phys.*, **35**, 2870-2876 (October 1964).
[97] S. E. Harris, "Stabilization and Modulation of Laser Oscillators by Internal Time-Varying Perturbation," *Appl. Opt.*, **5**, 1639-1651; and *Proc. IEEE*, **54**, 1401-1413 (October 1966).
[98] E. O. Ammann, B. J. McMurtry, and M. K. Oshman, "Detailed Experiments on He-Ne FM Lasers," *IEEE J. Quant. Elect.*, **QE-1**, 263-272 (September 1965).
[99] S. E. Harris and O. P. McDuff, "Theory of FM Laser Oscillation," *IEEE J. Quant. Elect.*, **QE-1**, 245-262 (September 1965).
[100] J. Owens, "Optical Refractive Index of Air: Dependence on Pressure, Temperature and Composition," *Appl. Opt.*, **6**, 51-95 (January 1967).
[101] W. R. C. Rowley and D. C. Wilson, "Wavelength Measurements of Helium-Neon Laser Emission," *J. Opt. Soc. Am.*, **56**, 259 (February 1966).
[102] E. Engelhard, "Wellenlängenstabilität eines Neon-Helium-Lasers," *Z. Angew. Phys.*, **20**, 404-407 (1966).
[103] K. D. Mielenz, K. F. Nefflen, K. E. Gillilland, R. B. Stephens, and R. B. Zipin, "Measurement of the 633-nm Wavelength of Helium-Neon Lasers," *Appl. Phys. Lett.*, **7**, 277-279 (1965).

APPENDIX

A

SELECTED LIST OF IMPORTANT GAS LASER TRANSITIONS

This list is abstracted from a total of some 1200 known gas laser wavelengths and represents those transitions which, in the author's opinion, are the most important and useful ones. Although such a classification is somewhat subjective, the following factors were considered in deciding that a transition should be included in the list: The transition is strong, well documented, and easily reproducible in its particular type of laser, and it is usable (or has been used) for applications other than laser research. Most of the lines listed here are obtainable from commercial equipment. When a line that does not meet our criteria is listed, it is because it has special interest, for example, because of very short or very long wavelength.

The references listed here will provide further information about these transitions; they are not necessarily the reports of initial discovery.

References to this table:

1. Reference 26 in bibliography. The appendix to this paper lists almost all gas laser transitions known at that time.
2. Reference 57 in bibliography. Much of this material is also contained in [56].
3. V. M. Kashlin and G. G. Petrash, *J.E.T.P. Lett.*, **3,** 55, (1966).
4. C. K. N. Patel, *Appl. Phys. Lett.*, **7**, 246 (1965).
5. J. D. Rigden, *Appl. Phys. Lett.*, **8,** 69 (1966).
6. M. Pollack et al., *Appl. Phys. Lett.*, **10,** 253 (1967).
7. W. M. Müller et al., *Appl. Phys. Lett.*, **10,** 93 (1967).
8. H. Steffen et al., *Phys. Lett.*, **20,** 20 (1966).

Selected List of Important Gas Laser Transitions

Wavelength (μ)	Spectrum	Species-Neutral, or Molecular, Ion	CW if So Listed, Otherwise Pulsed Only	Reference Number
0.235800	Ne IV	Ion		2
0.262490	Ar IV	Ion		2
0.332379	Ne II	Ion	CW	2
0.3371	N_2	Molecular		3
0.337833	Ne II	Ion	CW	2
0.339286	Ne II	Ion	CW	2
0.350742	Kr III	Ion		2
0.351113	Ar III	Ion		2
0.437073	Ar II	Ion	CW	2
0.454504	Ar II	Ion	CW	2
0.457720	Kr II	Ion	CW	2
0.457936	Ar II	Ion	CW	2
0.460302	Xe II	Ion	CW	2
0.461917	Kr II	Ion	CW	2
0.465795	Ar II	Ion	CW	2
0.468045	Kr II	Ion	CW	2
0.472689	Ar II	Ion	CW	2
0.476244	Kr II	Ion	CW	2
0.476488	Ar II	Ion	CW	2
0.476571	Kr II	Ion	CW	2
0.482518	Kr II	Ion	CW	2
0.484666	Kr II	Ion	CW	2
0.487986	Ar II	Ion	CW	2
0.49651	Ar II	Ion	CW	2
0.501717	Ar II	Ion	CW	2
0.504489	Xe II	Ion	CW	2
0.514533	Ar II	Ion	CW	2
0.520832	Kr II	Ion	CW	2
0.521790	Cl II	Ion	CW	2
0.52615	Xe II	Ion	CW	2
0.5287	Ar II	Ion	CW	2
0.530868	Kr II	Ion	CW	2
0.539215	Cl II	Ion	CW	2
0.5407	I II	Ion		2
0.541916	Xe II	Ion	CW	2
0.559237	O III	Ion		2
0.56772	Hg II	Ion		2
0.567956	N II	Ion		2
0.568192	Kr II	Ion	CW	2
0.5760	I II	Ion		2

Wavelength (μ)	Spectrum	Species-Neutral, or Molecular, Ion	CW if So Listed, otherwise Pulsed Only	Reference Number
0.59393	Ne	Neutral	CW	1
0.597112	Xe II	Ion	CW	2
0.60461	Ne	Neutral	CW	1
0.61180	Ne	Neutral	CW	1
0.6127	I II	Ion		2
0.614950	Hg II	Ion		2
0.627090	Xe II	Ion	CW	2
0.62937	Ne	Neutral	CW	[a]
0.63282	Ne	Neutral	CW	1
0.63518	Ne	Neutral	CW	1
0.64011	Ne	Neutral	CW	1
0.64710	Kr II	Ion	CW	2
0.65700	Kr II	Ion	CW	2
0.676457	Kr II	Ion	CW	2
0.687096	Kr II	Ion	CW	2
0.73048	Ne	Neutral	CW	1
0.79930	Kr II	Ion		2
0.8446	O	Neutral	CW	1 [b]
1.0798	Ne	Neutral	CW	1
1.0844	Ne	Neutral	CW	1
1.1143	Ne	Neutral	CW	1
1.1177	Ne	Neutral	CW	1
1.1390	Ne	Neutral	CW	1
1.1409	Ne	Neutral	CW	1
1.15228	Ne	Neutral	CW	1
1.15250	Ne	Neutral	CW	1
1.1601	Ne	Neutral	CW	1
1.1614	Ne	Neutral	CW	1
1.1767	Ne	Neutral	CW	1
1.1789	Ne	Neutral	CW	1
1.1985	Ne	Neutral	CW	1
1.2066	Ne	Neutral	CW	1
1.5231	Ne	Neutral	CW	1
2.0262	Xe	Neutral	CW	1
2.6511	Xe	Neutral	CW	1
3.3903	Ne	Neutral	CW	1
3.3913	Ne	Neutral	CW	1
3.5070	Xe	Neutral	CW	1

[a] Not listed in [1] by oversight. See 0.61180 μ.

[b] This line has a detailed fine structure, only the mean wavelength is listed.

Wavelength (μ)	Spectrum	Species-Neutral, or Molecular, Ion	CW if So Listed, otherwise Pulsed Only	Reference Number
5.6	CO	Molecular	CW	4[c]
10.5135	CO_2	Molecular	CW	1
10.5326	CO_2	Molecular	CW	1
10.5518	CO_2	Molecular	CW	1
10.5713	CO_2	Molecular	CW	1
10.5912	CO_2	Molecular	CW	1
10.6	CO_2	Molecular	CW	5[c]
10.6118	CO_2	Molecular	CW	1
10.6324	CO_2	Molecular	CW	1
10.6748	CO_2	Molecular	CW	1
10.6965	CO_2	Molecular	CW	1
10.7194	CO_2	Molecular	CW	1
10.7415	CO_2	Molecular	CW	1
10.7748	CO_2	Molecular	CW	1
10.7880	CO_2	Molecular	CW	1
27.974	H_2O	Molecular	CW	6
47.693	H_2O	Molecular	CW	6
78.455	H_2O	Molecular	CW	6
118.65	H_2O	Molecular	CW	7
220.34	H_2O	Molecular	CW	7
310.8870	HCN	Molecular	CW	7
336.5578	HCN	Molecular	CW	7
773.5	ICN	Molecular		8

[c] Listed wavelength is approximate mean of over 100 wavelengths not listed separately in the references.

APPENDIX B

EXACT WAVELENGTH OF STABILIZED HELIUM-NEON LASER

Best measured vacuum wavelength: 6329.9138 Å. Wavelength in dry air at 20°C, 760 torr containing 0.03% CO_2: 6328.1938 Å.

The following tables list corrections $\Delta\lambda$ to be added to the given air wavelength for other atmospheric conditions. To fair approximation, these corrections are additive. For further information, the reader is directed to [100] of the Bibliography.

TABLE B1
Pressure Correction at 20°C

Pressure (torr)	$\Delta\lambda$ (Å)
620	+0.3169
640	+0.2717
660	+0.2264
680	+0.1811
700	+0.1359
720	+0.0906
740	+0.0453

TABLE B2

Temperature correction at 760 torr

Temperature (°C)	$\Delta\lambda$ (Å)
15	−0.0299
20	0
25	+0.0290
30	+0.0569

TABLE B3

Correction at One Hundred Percent Humidity
(Actual correction is directly proportional to humidity)

Temperature (°C)	$\Delta\lambda$ (Å)
15	+0.0046
20	+0.0062
25	+0.0082
30	+0.0108

AUTHOR INDEX

SUBJECT INDEX